SUSTAINABLE DEVELOPMENT

A Geospatial and Statistical Perspective

AAP Research Notes on Optimization and Decision-Making Theories

SUSTAINABLE DEVELOPMENT

A Geospatial and Statistical Perspective

Qazi Mazhar Ali, DPhil (Oxon.)
Rizwan Ahmad, PhD
Irfan Ali, MPhil, PhD
Prasenjit Chatterjee, PhD
Editors

AAP | APPLE ACADEMIC PRESS

First edition published 2026

Apple Academic Press Inc.
1265 Goldenrod Circle, NE,
Palm Bay, FL 32905 USA

760 Laurentian Drive, Unit 19,
Burlington, ON L7N 0A4, Canada

CRC Press
2385 NW Executive Center Drive,
Suite 320, Boca Raton FL 33431

4 Park Square, Milton Park,
Abingdon, Oxon, OX14 4RN, UK

Apple Academic Press exclusively co-publishes with CRC Press, an imprint of Taylor & Francis Group, LLC

Library and Archives Canada Cataloguing in Publication

CIP data on file with Canada Library and Archives

Library of Congress Cataloging-in-Publication Data

CIP data on file with US Library of Congress

ISBN: 978-1-77964-323-0 (hbk)
ISBN: 978-1-77964-324-7 (ebk)
DOI: 10.1201/9781779643247

DEDICATION

The Editors would like to dedicate this book to their parents, life partners, children, students, scholars, friends, and colleagues.

ABOUT THE BOOK SERIES: AAP RESEARCH NOTES ON OPTIMIZATION AND DECISION MAKING THEORIES

Most real-world search and optimization problems naturally involve multiple criteria as objectives. Different solutions may produce trade-offs (conflicting scenarios) among different objectives. A solution that is better with respect to one objective may be a compromising one for other objectives. This compels one to choose a solution that is optimal with respect to only one objective. Due to such constraints, multi-objective optimization problems (MOPs) are difficult to solve since the objectives usually conflict with each other. It is usually hard to find an optimal solution that satisfies all objectives from the mathematical point of view. In addition, it is quite common that the criteria of real-world MOPs encompass uncertain information, which becomes quite a challenging task for a decision maker to select the criteria. Also, the complexities involved in designing mathematical models increase. Considering, planning, and appropriate decision-making require the use of analytical methods that examine trade-offs; consider multiple scientific, political, economic, ecological, and social dimensions; and reduce possible conflicts in an optimizing framework. Among all these, real-world multi-criteria decision-making (MCDM) problems related to engineering optimizations are categorically important and are quite often encountered with a wide range of applicability.

MCDM problems are basically a fundamental issue in various fields, including applied mathematics, computer science, engineering, management, and operations research. MCDM models provide a useful way for modeling various real-world problems and are extensively used in many different types of systems, including, but not limited to, communications, mechanics, electronics, manufacturing, business management, logistics, supply chain, energy, urban development, waste management, and so forth.

In the aforementioned cases, modeling of multiple criteria problems often becomes more complex if the associated parameters are uncertain and imprecise in nature. Impreciseness or uncertainty exists within the parameters due to imperfect knowledge of information, measurement uncertainty, sampling uncertainty, mathematical modeling uncertainty, etc. Theories like probability theory, fuzzy set theory, type-2 fuzzy set theory, rough set, grey theory, neutrosophic uncertainty theory available in the existing literature deal with such uncertainties. Nevertheless, the uncertain multi-criteria characteristics in such problems are not explored in depth, and a lot can be achieved in this direction. Hence, different mathematical models of real-life multi-criteria optimization problems can be developed on various uncertain frameworks with special emphasis on sustainability, manufacturing, communications, biomedical, electronics, materials, energy, agriculture, environmental engineering, strategic management, flood risk management, supply chain, waste management, transportations, economics, and industrial engineering problems, to name a few.

This series reports on current trends and advances in optimization and decision-making theories in a wider range of domains for academic and research institutes along with industrial organizations.

Features of the volumes include recent trends, model extensions, developments, real-time examples, case studies, and applications. The volumes aim to serve as valuable resources for undergraduate, postgraduate, and doctoral students, as well as for researchers and professionals working in a wider range of areas.

CURRENT & FORTHCOMING BOOKS IN THE SERIES

- **Multi-Criteria Decision-Making Techniques in Waste Management: A Case Study of India**
  Editors: Suchismita Satapathy, Debesh Mishra, and Prasanjit Chatterjee
- **Applications of Artificial Intelligence in Business and Finance: Modern Trends**
  Editors: Vikas Garg, Shalini Aggarwal, Pooja Tiwari, and Prasenjit Chatterjee
- **Advanced Computer Science Applications: Recent Trends in AI, Machine Learning, and Network Security**
  Editors: Karan Singh, Latha Banda, and Manisha Manjul
- **Decision-Making Models and Applications in Manufacturing Environments**
  Editors: Pushpdant Jain, Kumar Abhishek, and Prasenjit Chatterjee
- **Precision Agriculture for Sustainability: Use of Smart Sensors, Actuators, and Decision Support Systems**
  Editors: Narendra Khatri, Ajay Kumar Vyas, Celestine Iwendi, and Prasenjit Chatterjee
- **Attacks on Artificial Intelligence: The New Facets of Cyber Ecospace**
  Editors: Kukatlapalli Pradeep Kumar, Boppuru Rudra, Vinay Jha Pillai, and Prasenjit Chatterjee
- **Computational Optimization, Modeling, and Simulation for Engineering Applications**
  Editors: Anupam Shukla, Sourabh Rungta, Mohan Awasthy, and Rakesh L. Himte
- **Advances in Intelligent Systems: Paradigms and Applications**
  Editors: Amit Krishna Dwivedi, Manisha Guduri, Rajanikanth Aluvalu, and Uma Maheshwari V.
- **Advances in Data Science and Computing Technology: Methodology and Applications**
  Editors: Suman Ghosal, Amitava Choudhury, Vikram Kumar Saxena, Arindam Biswas, and Prasenjit Chatterjee
- **Multimedia Security: Tools, Techniques, and Applications**
  Editors: Bhaskar Mondal and Shyam Singh Rajput
- **Sustainable Development: A Geospatial and Statistical Perspective**
  Editors: Qazi Mazhar Ali, Rizwan Ahmad, Irfan Ali, and Prasenjit Chatterjee
- **Advances in Intelligent Systems**
  Editors: Manisha Guduri, Uma Maheswari V, Rajanikanth Aluvalu, and Amit Krishna Dwivedi

- **Machine Learning in Biomedical and Health Informatics: Current Applications and Challenges**
  Editors: Sudip Kumar Sahana, Rajendrani Mukherjee, Panchali Datta Choudhury, and Prasenjit Chatterjee

ABOUT THE EDITORS

Qazi Mazhar Ali, DPhil (Oxon.)
Former Dean Faculty of Science and Professor, Department of Statistics & Operations Research, Aligarh Muslim University, Aligarh, India

Qazi Mazhar Ali, the former Dean of the Faculty of Science and Professor of Statistics at the *Department of Statistics & Operations Research, Aligarh Muslim University, Aligarh, India, earned his doctorate degree from the University of Oxford in 1993.* His research interest includes image data analysis. Recently, he has been designated as the primary point of contact for the United Nations Global Geospatial Information Management (UN-GGIM), and he is actively engaged and focusing on Sustainable Development Goal 2030 through a statistical data network. Within his broad area, he has published his work in digital image analysis, classification (remotely sensed images), epidemiology, and bioinformatics in UK and Germany. His epidemiology paper appeared in the *Journal of Helminthology* (UK). In terms of innovation in teaching, he contributed in popularizing PC-based computer education in the 1980s and continues to advance it still. Additionally, he contributed in the efforts to revive the instructions for the award of degrees in remote sensing at AMU. He supervised several MSc projects on Sustainable Development Goals.

In the past, he taught courses on various topics, including simulation, computer programming, demography, statistics for business studies, biostatistics, etc. Furthermore, he had been actively involved in the academic administration through various posts such as Director Computer Centre (AMU), He served as Chairperson of the Department of Statistics & Operations Research and that of Interdisciplinary Department of Remote Sensing & GIS Applications, AMU.

Rizwan Ahmad, PhD
Faculty, Interdisciplinary Department of Remote Sensing & GIS Applications, Aligarh Muslim University, Aligarh, India

His responsibilities include teaching courses of MSc and PhD programs in Remote Sensing & GIS Applications, AMU, Aligarh. The Research Interests of Dr Rizwan Ahmad include Earth observation, Geospatial technology applications, and GIS data integration. He has a doctorate in Geoscience and vast research & teaching experience in India and abroad. He has worked with prestigious institutes like IIRS-ISRO, Forest Survey of India (MoEF & CC, Govt. of India), Dehradun and Eritrea Institute of Technology (EIT), Asmara, Eritrea (North East Africa) as Assistant Professor of Geospatial technology. He also had an opportunity to work in ISRO (Respond) project. Dr Ahmad specializes in Remote sensing & GIS applications and Applied geology.

Dr Rizwan has also completed many certified Training Programs of NASA's Applied Remote Sensing during 2020 - 2023, including Citizen science connecting Remote sensing, Assessing data from Geostationary satellite, GEE for land monitoring, SDG monitoring, Hyperspectral data, UHI, WQ, and Species distribution. His major publications are in applied geoscience using geospatial techniques; Dr Rizwan worked as Knowledge Partner in "UN-SDGs Goals 2030" using geospatial and statistical data in LK Singhania Education Center (participating in UNESCO asp.net), Rajasthan. This University designated him as the second point of contact for the United Nations – Global Geospatial Information Management Academic Network (UN GGIM) in 2020. Dr Rizwan has the honor of embedding high-resolution satellite data from the university campus in the historical event of "TIME CAPSULE" on the occasion of the University Bicentenary Celebration 2021. He has served as a reviewer in many journals of repute, in which IEEE—Journal of Selected Topics in Applied Earth Observations & Remote Sensing, Remote Sensing Letters, Journal of Geological Society of India (JGSI), Frontiers in Earth Sciences, Geocarto International, International *Journal of Environmental Research, Cogent Food & Agriculture Journal* by Taylor & Francis and *Disaster Advances* are few of them. He is a member of the Editorial Board of CRC Press and is also associated with Springer Publications to review the preliminary information of several book proposals. Dr. Rizwan has also served in various university administrative positions like Assistant Superintendent of Examinations, Examiners, Scrutinizer, Verifying officer, and Tabulator, etc., from time to time.

Irfan Ali, MPhil, PhD
Senior Assistant Professor, Department of Statistics and Operations Research, Aligarh Muslim University, Aligarh, India

Dr. Irfan Ali received an MPhil. and PhD degree in statistics from Aligarh Muslim University. He is currently working as a Senior Assistant Professor at the Department of Statistics and Operations Research at Aligarh Muslim University. In 2024, one of the world-renowned scholar agencies, Scholar GPS, recognized Dr. Ali's contributions to the Mathematical Optimization and Environment statistics fields and ranked him among the world's top 0.5% of scholars. He received the Post Graduate Merit Scholarship Award during his MSc (statistics) and the UGC-BSR Scholarship awarded during his PhD (statistics) programs. His research interests include Applied statistics, mathematical optimization, fuzzy logic and optimization, multiobjective optimization, supply chain management, UN Sustainable Development Goals and Renewable Energy Production modelling. He has supervised MSc, projects, MPhil, dissertations and PhD thesis under his supervision in mathematical optimization and applied statistics topics. He has completed research project UGC–Start-Up Grant Project, UGC, New Delhi, India and the Mitacs Globalink Research Award for a joint research program with the University of Regina, Canada. He has published over 105 research articles in the Scopus-indexed database and 71 in the Web of Science database. His Google citations are 2256 with a 26-h index and 63-i10 index. He has published six edited books for Taylor & Francis and Springer Nature publishers, and some are in the process of publication. All these books have been indexed in one of Elsevier's reputed Scopus indexing databases. He has also published one textbook, "Optimization-with-LINGO-18-Problems-and-Applications". This book is helpful for academicians, practitioners, students, and researchers in OR. He is a Lifetime Member of various professional societies: Operational Research Society of India, Indian Society for Probability and Statistics, Indian Mathematical Society, and The Indian Science Congress Association. He is an associate EC Member in the Indian Society for Probability and Statistics (ISPS) [2021, 2022, 2023, 2024]. He has delivered invited talks at several universities and Institutions. He also serves as associate editor/editor/editorial board member of more than 10 journals of international reputation.

Prasenjit Chatterjee, PhD
Professor of Mechanical Engineering and Dean (Research and Consultancy), MCKV Institute of Engineering, West Bengal, India

Prasenjit Chatterjee, PhD, is currently Dean (Research and Consultancy) at the MCKV Institute of Engineering, West Bengal, India. He has over 6400 citation and has published over 130 research papers in various international journals and peer-reviewed conferences. He has authored and edited more than 15 books on intelligent decision-making, supply chain management, optimization techniques, risk, and sustainability modeling. He has received numerous awards, including best track paper awards, outstanding reviewer awards, best paper awards, outstanding researcher awards, and university gold medals. He has been a guest editor of several special issues of various indexed journals. He has authored and edited several books on decision analysis, disruptive technologies, intelligent computing, supply chains, and sustainability modeling. He is the Lead Series Editor of the book series *Disruptive Technologies and Digital Transformations for Society 5.0*. He is the Founder and Lead Series Editor of the book series *Concise Introductions to AI and Data Science*; *AAP Research Notes on Optimization and Decision-Making Theories*; *Frontiers of Mechanical and Industrial Engineering; Smart and Intelligent Computing in Engineering*. Dr. Chatterjee is one of the developers of two multiple-criteria decision-making methods: Measurement of Alternatives and Ranking according to COmpromise Solution (MARCOS) and Ranking of Alternatives through Functional mapping of criterion sub-intervals into a Single Interval (RAFSI).

CONTENTS

CONTRIBUTORS

Stephy Abraham
Interdisciplinary Department of Remote Sensing & GIS Applications, Aligarh Muslim University, Aligarh, Uttar Pradesh, India

Rizwan Ahmad
Interdisciplinary Department of Remote Sensing & GIS Applications, Aligarh Muslim University, Aligarh, Uttar Pradesh, India

M. A. Ahmed
Department of Agricultural Economics & Extension, Modibbo Adama University, Yola Nigeria

Uzma Ajmal
Department of Geography, Aligarh Muslim University, Aligarh, Uttar Pradesh, India

S. K. Ajim Ali
Department of Geography, Aligarh Muslim University, Aligarh, Uttar Pradesh, India

Irfan Ali
Department of Statistics & Operations Research, Aligarh Muslim University, Aligarh, Uttar Pradesh, India

Qazi Mazhar Ali
Department of Statistics and Operations Research, Aligarh Muslim University, Aligarh, Uttar Pradesh, India

Aqsa Anjum
Department of Commerce, Aligarh Muslim University, Aligarh, Uttar Pradesh, India

Ziaur Rehman Ansari
Interdisciplinary Department of Remote Sensing & GIS Applications, Aligarh Muslim University, Aligarh, Uttar Pradesh, India

Tanveer Asif
Department of Geography, Aligarh Muslim University, Aligarh, Uttar Pradesh, India

Pankaj Chauhan
Wadia Institute of Himalyan Geology, Dehradun

Jahangir Chauhan
Department of Commerce, Aligarh Muslim University, Aligarh, Uttar Pradesh, India

O. U. Cynthia
Department of Mathematics and Statistics, University of Port Harcourt, Nigeria

Samreen Fatima
Interdisciplinary Department of Remote Sensing & GIS Application, Aligarh Muslim University, Aligarh, Uttar Pradesh, India

Rounaq Goenka
SC National Atmospheric Research Laboratory, National Remote Sensing Centre, Department of ISRO, Government of India Balanagar, Hyderabad, Andhra Pradesh, India

Abhishek Gupta
Interdisciplinary Department of Remote Sensing and GIS Application, Aligarh Muslim University, Aligarh, Uttar Pradesh, India

Asif Iqubal
Interdisciplinary Department of Remote Sensing and GIS Applications, Aligarh Muslim University, Aligarh, Uttar Pradesh, India

Mohammed Irshad
International Crops Research institute for the Semi-Arid Tropics, Patancheru, Hyderabad, Andhra Pradesh, India

Saleha Jamal
Department of Geography, Aligarh Muslim University, Aligarh, Uttar Pradesh, India

Akram Javed
Department of Geology, Aligarh Muslim University, Aligarh, Uttar Pradesh, India

Niraj Prakash Joshi
Center for Peaceful and Sustainable Future, the IDEC Institute/ TAOYAKA Program/ International Economic Development Program, Graduate School of Humanities and Social Sciences/ Social Innovation Science, Graduate School of Innovation and Practice for Smart Society, Hiroshima University, Hiroshima, Japan

Waiza Khalid
Interdisciplinary Department of Remote Sensing & GIS Application, Aligarh Muslim University, Aligarh, Uttar Pradesh, India

Haris Hasan Khan
Interdisciplinary Department of Remote Sensing and GIS Applications, Aligarh Muslim University, Aligarh, Uttar Pradesh, India

Mohammad Muqeet Khan
Department of Geography, Jamia Millia Islamia University, New Delhi, India

Arina Khan
Department of Geological Sciences,George Washington University, Washington, D.C., USA

Taqveem Ali Khan
Department of Geology, Aligarh Muslim University, Aligarh, Uttar Pradesh, India

Haris Hasan Khan
Interdisciplinary Department of Remote Sensing and GIS Applications, Aligarh Muslim University, Aligarh, Uttar Pradesh, India

Zeba Khanam
Post Doctoral Fellow (ICSSR), Jawaharlal Nehru University, New Delhi

Tanzeel Khatoon
Interdisciplinary Department of Remote Sensing & GIS Applications, Aligarh Muslim University, Aligarh, Uttar Pradesh, India

Ramesh Krishnamurthy
Wildlife Institute of India Dehradun, Uttarakhand

Farheen Mahtab
Interdisciplinary Department of Remote Sensing & GIS Applications, Aligarh Muslim University, Aligarh, Uttar Pradesh, India

M. Mijinyawa
Department of Statistics & Operations Research, Modibbo Adama University, Yola, Nigeria

U. M. Modibbo
Department of Operations Research, Modibbo Adama University Yola, Nigeria

Aswathy N. S.
Geological Survey of India Western Region, Jaipur, Rajasthan, India

Muhammad Zaid Qamar
Interdisciplinary Department of Remote Sensing & GIS Applications, Aligarh Muslim University, Aligarh, Uttar Pradesh, India

P. S. Roy
International Crops Research institute for the Semi-Arid Tropics, Patancheru, Hyderabad, Andhra Pradesh, India

Kalachand Sain
Wadia Institute of Himalayan Geology, Dehradun, India

Mohammad Sajid
Interdisciplinary Department of Remote Sensing & GIS Applications, Aligarh Muslim University, Aligarh, Uttar Pradesh, India

Anand Sharma
Regional Meteorological Centre, Indian Meteorological Department (IMD), Lodi Road, New Delhi

Ateebur Rehman Sherwani
Department of Statistics and Operations Research, Aligarh Muslim University, Aligarh, Uttar Pradesh, India

Mansoor Alam Siddiqui
Department of Geography, Women's College, Aligarh Muslim University, Aligarh, Uttar Pradesh, India

Ghazala Siraj
Department of Geology, Aligarh Muslim University, Aligarh, Uttar Pradesh, India

Sameer Tiwari
Wadia Institute of Himalayan Geology, Dehradun, India

Alia Yusuf
Geology Section, Women's College, Aligarh Muslim University, Aligarh, Uttar Pradesh, India

Qazi Zeeshan Ahmad Jalali
GIS Data Analyst at GAIT Global

ABBREVIATIONS

AIC	Akaike information criterion
ANN	artificial neural network
AOI	area of interest
CFI	composite functional index
CHS	composite hierarchical score
CMS	composite market-facilities score
CPCB	Central Pollution Control Board
CRB	crop residue burning
DEM	digital elevation model
ECC	electricity consumption
FEM	finite element method
GCP	ground control points
GHG	greenhouse gas
GIS	geographic information system
GP	goal programming
GPS	Global Positioning System
GWR	geographically weighted regression
HRU	hydrologic response units
HS	household assets
IDW	Inverse Distance Weighted
IWRIS	India Water Resource Information System
KWA	kinematic wave analogy
LST	land surface temperature
LULC	land use and land cover
MDGs	millennium development goals
MRR	micro rain radar
NDBI	normalized difference built-up index
NDVI	normalized difference vegetation index
NIR	near infrared
OLI	operation land imager
OR	ordinary kriging
PCA	principal component analysis
PET	potential evapotranspiration

PM	particulate matter
PV	photovoltaic
QoL	quality of life
RK	regression kriging
SDGs	sustainable development goals
SE	standard error
SPV	solar photovoltaic
SRTM	Shuttle Radar Topography Mission
ST	surface temperature
SWIR	shortwave infrared
TIRS	thermal infrared sensor
TWI	topographic wetness index
UHI	urban heat island
WSR	weighted spatial regression

ACKNOWLEDGMENTS

This acknowledgment expresses gratitude to the individuals who played significant roles in making the book a reality. The authors receive appreciation for their contributions and dedication to the chapters, enriching the book with their expertise. The reviewers are acknowledged for their valuable input, enhancing the quality, coherence, and presentation of the content.

Lastly, the Apple Academic Press team, particularly Ashish, Sandra, and Rakesh, are recognized for their motivation and support.

The editors express sincere thanks to everyone involved in bringing this project to fruition.

—Qazi Mazhar Ali, DPhil (Oxon.)
Rizwan Ahmad, PhD
Irfan Ali, MPhil, PhD
Prasenjit Chatterjee, PhD

PREFACE

The United Nations introduced the Sustainable Development Goals (SDGs) in 2015 as a global rallying point, aiming to eradicate poverty, safeguard the environment, and secure collective prosperity by 2030. This framework encompasses 17 diverse objectives, spanning from the eradication of poverty and hunger to the advancement of health, education, gender parity, access to clean water, and the promotion of sustainable energy sources. The overarching ambition of these goals is to comprehensively address the various facets of worldwide development, endeavoring to establish a more sustainable and fairer world by the designated year of 2030. These goals serve not just as aspirations but as guiding principles for policymakers, governments, international organizations, businesses, and communities worldwide. They inspire action and inform decision-making, steering resource allocation, policy formulation, and developmental strategies toward a collective vision of a better, more sustainable world. Geospatial perspectives play a significant role in achieving these goals by providing critical spatial information, analysis, and tools that aid in understanding and addressing various global challenges and to derive insights from mapping, satellite imagery, and spatial analysis.

Thus, the main objective of this book is to provide a comprehensive exploration of the intersection between SDGs and geospatial technology. The book aims to analyze how geospatial and statistical information, including satellite imagery, Geographic Information System (GIS), remote sensing, statistical tools, and data analysis techniques effectively employed to assess, monitor, and facilitate sustainable development initiatives. Through an interdisciplinary approach, the book intends to highlight the crucial role of geospatial data in understanding, monitoring, and addressing global challenges and other key facets of sustainable development. This book consists of 14 chapters as discussed below:

Chapter 1 endeavors to advance the SDGs, underscoring the necessity of robust decision-making tools, particularly optimization modeling, to effectively attain these objectives. It proposes a multicriteria approach employing a mathematical model aimed at enhancing energy efficiency, bolstering economic productivity, and curbing greenhouse gas emissions.

Chapter 2 explores how changes in land use affect water flow and flood potential in river basins. It specifically examines the porandalar basin, utilizing a hydrological model to quantify flood potential changes across different land use scenarios from 1973 to 2013 and a projected future scenario.

Land Use and Land Cover (LULC) changes play a vital role in various applications, influencing land planning and climate patterns. Chapter 3 examines LULC alterations in Raichur district, Karnataka, India, using Remote Sensing and GIS. Landsat 8 data from December 2015 underwent supervised classification to understand LULC of the considered region. Employing the maximum likelihood method with ground control points from Google Earth Pro, the study analyzes the six bands of Landsat imagery.

Chapter 4 analyzes the air quality specifically during October and November from 2017 to 2019 in some selected Indian regions. Daily measurements of particulate matter, ozone, nitrogen dioxide, and carbon monoxide concentrations are collected. Averaging these levels and employing interpolation techniques, the chapter presents an overview of how these pollutants impact the regions, especially during periods of heightened pollution.

Water table elevation maps are essential for hydrogeological studies, aiding in understanding groundwater flow, recharge areas, and the impact of human and natural influences. These maps are traditionally created by gathering water level data from wells and using interpolation to create a representative surface. Chapter 5 evaluates different interpolation techniques for groundwater elevation data. The goal is to identify the most effective technique for groundwater agencies to produce more accurate water table maps.

Chapter 6 identifies four significant contributors to subsidence accounting for spatial variability in coal-mining activities including coal depth, surface slope, inclination, and thickness. The study also juxtaposed the subsidence map, generated using geographically weighted regression (GWR) method with these four factors as explanatory variables, to assess and estimate subsidence in unmapped areas affected by alterations in artificial landforms.

Chapter 7 highlights a significant correlation between sanitation subindex and the household latrine indicator with the Quality of Life (QoL) index, underscoring the vital role of sanitation in QoL. Moreover, specific dimensions within the living standards exhibit strong correlations and variance with different sub-indexes.

Chapter 8 investigates the evolution of LULC alongside climatic variations, recognizing this as a prominent and impactful aspect of global environmental change. The study investigates using meteorological data encompassing five key climatic parameters such as temperature, rainfall, wind velocity, sunshine hours, and relative humidity.

Chapter 9 focuses on comparing ground-based precipitation data with information obtained from satellite-derived precipitation products.

Chapter 10 focuses on analyzing the Urban Heat Island phenomenon using satellite imagery. Remote sensing data was employed to calculate Land Surface Temperature, Normalized Difference Vegetation Index, and Normalized Difference Built-up Index for the respective years.

Chapter 11 is dedicated to constructing an effective model and solution process utilizing goal programming. This model aims to simultaneously optimize various crucial factors such as electricity consumption, economic growth, greenhouse gas emissions, and workforce allocation. The study utilized LINGO 17.0, a modeling and optimization tool, to ensure the accuracy and applicability of the results.

In Chapter 12, the focus is on evaluating whether transitioning to solar power generation can lead to a reduction in overall costs, particularly in electricity expenses for educational institutions. The study likely explores the potential economic benefits and feasibility of adopting solar energy as an alternative or supplementary source for these institutions. This investigation could encompass factors such as initial setup costs, operational expenses, and the long-term financial implications of such a transition.

Chapter 13 is dedicated to identifying the hierarchy of specific rural markets and examining the connection between market hierarchy and the development of infrastructure and service facilities. The study involves calculating the composite percent change using 20 selected variables that measure the changes in market infrastructure and service facilities over a 20-year period.

In Chapter 14, the focus lies on quantifying potential evapotranspiration (ETo) in glacier valleys. The study involves computing ETo utilizing *in-situ* meteorological data and the Penman–Monteith equation. Two standard methods, the FAO method and Cropwat 8.0, were employed to calculate ETo. The study then compared and validated the results obtained from both methods by using modeled GLCF Modis data. This research aims to analyze and verify the accuracy of different methods in calculating potential evapotranspiration, specifically in glacier valleys.

—Editors

CHAPTER 1

A MULTICRITERIA OPTIMIZATION MODELING FOR THE SUSTAINABLE DEVELOPMENT GOALS OF NIGERIA: MISSION 2030

U. M. MODIBBO[1], M. MIJINYAWA[2], M. A. AHMED[3], and O. U. CYNTHIA[4]

[1]*Department of Operations Research, Modibbo Adama University Yola, Nigeria*

[2]*Department of Statistics & Operations Research, Modibbo Adama University, Yola, Nigeria*

[3]*Department of Agricultural Economics & Extension, Modibbo Adama University, Yola, Nigeria*

[4]*Department of Mathematics and Statistics, University of Port Harcourt, Nigeria*

ABSTRACT

The ever-increasing demand for attaining sustainable development goals (SDGs) by all United Nations member countries cannot be overemphasized. Decision-makers (DM) and policy designers are looking for rigorous and robust DM tools, such as optimization modeling, to reap the SDGs targets economically and efficiently. This study investigates Nigeria's SDGs achievement through a multicriteria approach. A mathematical model developed and discussed adequate strategies that simultaneously address efficiency in

Sustainable Development: A Geospatial and Statistical Perspective. Qazi Mazhar Ali, Rizwan Ahmad, Irfan Ali, & Prasenjit Chatterjee (Eds.)

DOI: 10.1201/9781779643247-1

energy consumption, economic productivity, and reduction in greenhouse gas emissions. With the COVID-19 disruption, the targeted goals in 2030 are not realistic. Therefore, the government must revisit various indicators and make the necessary adjustments to realize the 2030 agenda mission.

1.1 INTRODUCTION

Regarding social, environmental, and economic factors, the world is facing a significant challenge in growth and development sustainability. In September 2015, the United Nations (UN) General Assembly came up with an agenda for 2030 known as the sustainable development goals (SDGs). Because of the difficulties encountered by the previous Millennium Development Goals (MDGs), the UN General Assembly committed itself to reducing severe poverty through international cooperation, and it set eight goals with 20 targets and 60 indicators to be completed within 15 years (2015). It is one of the most profound commitments in developing nations' history. The goals, according to Kamepalli and Pattanayak (2015), are to "reduce child mortality, enhance maternal health, fight HIV/AIDS, malaria, and other diseases; obliterate extreme poverty and hunger; make elementary education accessible to everyone; advance gender equality and women's empowerment; create a global development collaboration and ensure environmental sustainability."

The SDG agenda comprises 17 goals, 169 targets, and 244 indicators. According to Modibbo et al. (2021a), the goals are the following:

1. Ending poverty at all levels.
2. Ending hunger via sufficient nutrition and food safety for everyone, and promoting sustainable agriculture.
3. Attainment of a healthy life for all in all age brackets.
4. Providing equitable and quality education and lifelong learning opportunities.
5. Gender equality empowers girls and women of all ages of governance.
6. For a sustainable future, everyone needs access to safe drinking water and sanitation.
7. Provide economical, sustainable, and dependable contemporary energy services.
8. Encourage robust, inclusive, and long-term economic growth as well as decent work for everybody.
9. Encourage sustainable industrial development.

10. Lessen inequality both domestically and internationally.
11. Creating inclusive, secure, and sustainable towns and cities.
12. Encourage sustainable production and consumption practices.
13. Encourage climate change initiatives at all levels.
14. Achieve marine resource conservation and sustainable ocean and sea use.
15. Preserve and restore terrestrial ecosystems, and put an end to all biodiversity loss.
16. Create inclusive societies that uphold the rule of law, establish effective institutions, and develop and improve the implementation tools and international collaboration to maintain and raise performance standards.
17. International collaboration for sustainable development.

According to the SDGs Report 2022, the 2030 Agenda for Sustainable Development is being threatened by cascading and connected problems for humanity's survival and is in grave danger. The apogee of the information is the confluence of problems; conflict, climate change, and COVID-19 are the primary issues influencing the SDGs, including peace and security, the environment, health, education, and food and nutrition. Additionally, it describes the reversal of decades of progress in reducing hunger and poverty, enhancing health and education, delivering crucial services, and much more. She also identifies areas that require immediate attention if the SDGs are to be saved and real progress made for people and the environment by 2030.

1.2 SOCIOECONOMY OF NIGERIA

Nigeria is the African continent's colossus, with a total land area of 923,768 square kilometers (356,667 square miles), 910,771 square kilometers (351,649 square miles) of which is land (98.6%). The projected population is over 200 million, with an annual growth rate of 2.60%. The country is projected to become the third-largest economy in Africa by the year 2050 and is currently ranked seventh in the world. Forty-six million people are middle class, and the country has the largest population compared to other African nations. Nigeria has the largest market in terms of nominal GDP among African countries post-GDP rebasing. Economically, it was projected to be the ninth largest country in the world by the year 2050 in terms of GDP purchasing power parity (Hawksworth and Cookson, 2006; Dadush and Stancil, 2010; Modibbo et al., 2021c).

The country is comprised of six zones geopolitically: north-central, northeast, north-west, south-east, South–South, and southwest. Furthermore, the country consists of 36 states and the Federal Capital Territory. Sambo (2008) reported that all states in Nigeria have their gifts of natural resources, ranging from vegetation cover suitable for agriculture to thick forests, rivers, lakes, and mountains found almost all over the nation. The geometric effects of population growth are tied to the continuous high demand for utilizing natural, human, capital, and material resources in Nigeria.

According to the National Bureau of Statistics, the GDP growth, as shown in Figure 1.1, depicts the level at which nations stand. This calls for interaction among the social, economic, environmental, energy, andinstitutional processes to guarantee the future sustainable development of the country.

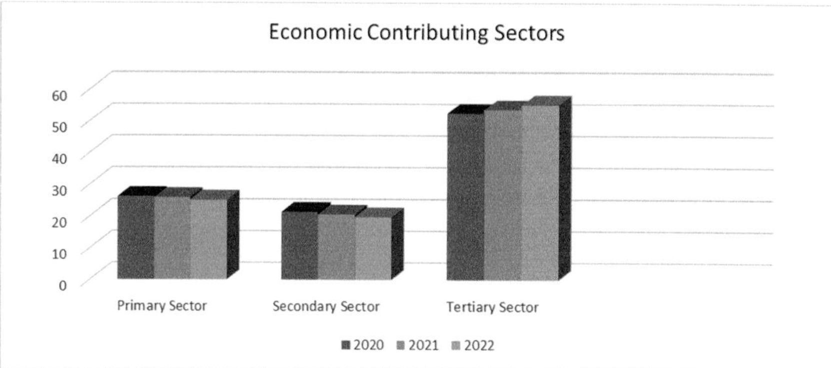

FIGURE 1.1 Nigerian average GDP percentage growth from the last quarter of 2020–2022.

According to the Nigeria Multidimensional Poverty Index 2022, more than a quarter of all potential deprivations are experienced by the poor in Nigeria. The North is home to 65% of the world's multidimensionally poor, or 86.1 million people, while the South is home to 35% of them, or roughly 47 million people.

A sizeable section of this underprivileged population lacks access to healthcare, safe cooking fuel, and proper sanitation. Most regions of sub-Saharan Africa, Nigeria inclusive, have not been doing as expected in most developmental agendas compared to other settings worldwide. On MDGs (2000–2015) alone, 72%, 68%, and 45% out of the 54 African countries were seriously off track on the improved situation, reduction of maternal mortality, and eradication of extreme poverty and hunger (World Bank, 2022).

Globally, it is not easy to sustain current social, environmental, and economic progress levels. The MDGs could not succeed due to challenges. The UN endorsed MDGs' failures, now known as the SDGs, Agenda 2030, in 2015. Global environmental deterioration and pollution are resulting from the exponential population increase and the rapidly rising demand for energy for production and consumption. After significant discussions with member nations, the UN Assembly developed 17 goals with more than 200 indicators in September 2015, with 2030 as the actualization year. The SDGs are interrelated and can be divided broadly into three main categories in this study: social, economic, and environmental goals.

Governments and other stakeholders must develop and put into effect feasible policies in each member nation if these sets of goals are to be accomplished. It is necessary to consume energy efficiently, enhance the economy, and reduce emissions to achieve eco-friendly environments. It is crucial to have policies addressing critical factors like employment opportunities, power availability, economic growth, and greenhouse gas (GHG) reduction. However, because of the uneven distribution of natural resources and the development of technology, these factors might not be the same in every country. The uneven distribution of resources and technological advances might be one of the causes of the varied ways that researchers and policymakers worldwide have applied optimization techniques to address these problems. The country's annual growth rate and population total are depicted in Figure 1.2, alongside the UN projections up to 2100 years.

1.3 SDGS OPTIMIZATION MODELING

An optimization is obtaining the best out of many possible best (optimal or near-optimal), where the goals are multiple and conflicting, getting the optimal solution is impossible. Hence, the term Pareto optimal or satisficing solution results in such cases. Several disciplines have been using this optimization concept to address real-life problems peculiar to those areas in engineering, management, social, pure, applied, biological, and environmental studies (Ali et al., 2021; Modibbo et al., 2021b). Therefore, SDGs, which encompass social, economic, and ecological components, are not exceptional in using optimization techniques to optimize the different functions depending on the goals and interests of the decision-makers (Jayaraman et al., 2017). Several studies applied preemptive goal programming (GP), fuzzy (GP), and GP with satisfaction functions. In contrast, others employed intuitionistic fuzzy GP and neutrosophic programming.

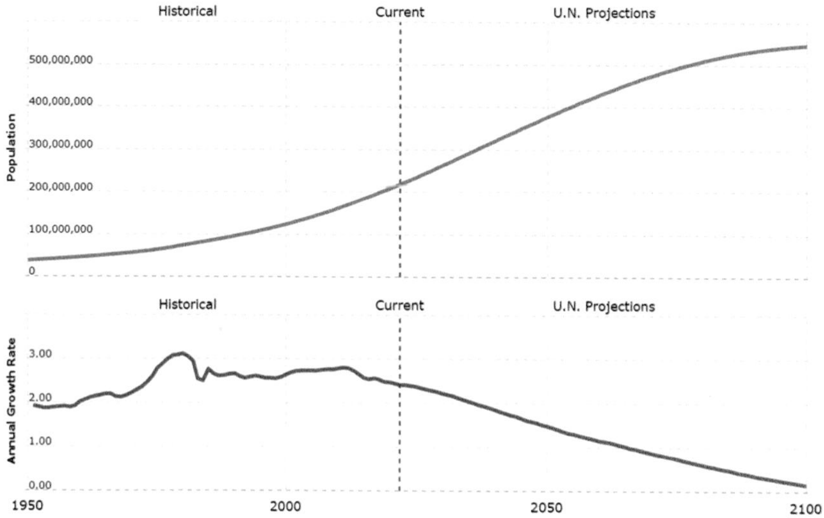

FIGURE 1.2 Nigerian population and growth rate projections.

1.3.1 THE SDGS MODEL

$$Z_1(X) = \sum_{i=1}^{3} [[(\text{GDP})_i]] x_j \tag{1.1}$$

$$Z_2(X) = \sum_{i=1}^{3} [(\text{ECC})_j] x_j \tag{1.2}$$

$$Z_3(X) - \sum_{i=1}^{3} [(\text{GHG})_i] x_j \tag{1.3}$$

Subject to the set of constraints:

$$\sum_{i=1}^{3} [(\text{GDP})_j] x_j \geq (\text{GDP})_g \tag{1.4}$$

$$\sum_{i=1}^{3}\left[(\mathrm{ECC})_j\right]x_j \geq (\mathrm{ECC})_g \quad (1.5)$$

$$\sum_{i=1}^{3}\left[(\mathrm{GHG})_j\right]x_j \leq (\mathrm{GHG})_g \quad (1.6)$$

$$\sum_{i=1}^{3} x_j \leq (\mathrm{EMP})_g \quad (1.7)$$

$$E_j \leq x_j \leq E_g \; j = 1,2,3 \quad (1.8)$$

1.4 RESULT ANALYSIS AND DISCUSSION

The above models (1.1)–(1.9) have been solved using the data in Tables 1.1 and 1.2 with a Lingo version 16.0, and the results are presented in Table 1.3. From the results, the outcome showed that D12, D22, and D32 had significant positive deviations. The values can be understood in this way: in order to concurrently satisfy all four criteria and be able to lower GHG emissions in Nigeria by 2030, a sizable amount of energy and GDP are needed. It implies that in order to meet Nigeria's rising energy needs, other renewable energy sources like wind, nuclear, and solar must be added. Only then will the country's Sustainable Goals Mission 2030 be realized.

TABLE 1.1 Target Goals by the Year 2030

Economic Indicator	GDP (Million NAIRA)	ECC (in Billion kWh)	GHG (Mt CO_2 eq)	Number of Employment (Million)
Value	44,601,361.13	29,011.42	308.180	69,675,468.12
Goal by the year 2030	6,699,450,000	54,895.58122	453	99,675,468
Annual growth rate	7.1%	6.71%	2.6%	0.70%

Sources: NBS (2022), CBN 2021 bulletin, and Modibbo et al. (2021a).

TABLE 1.2 Current Values for the Three Economic Sectors

Economic Sectors	GDP (Million NAIRA)	ECC (in kWh)	GHG (in Mt CO_2)	Employment ('000)
Primary	14,395,264.42	5222.056	55,472.4	20,902,640
Secondary	8,792,476.85	12,184.80	129,435.6	13,935,094
Tertiary	21,413,619.86	11,604.60	123,272	34,837,734

Source: From the authors computation.

TABLE 1.3 Lingo Solver Result Version 16.0

Deviations/Variable	Value	Reduced Cost
D12	50,142,860,000	0.000000
D22	65,168,930,000	0.000000
D32	14,200,110,000	0.000000
D11	0.000000	0.2000000
D21	0.000000	0.2000000
D31	0.000000	0.2000000
X1	209,026.4	0.000000
X2	39,350.9	0.000000
X3	348,377.3	0.000000

1.5 CONCLUSION

In developing policies and addressing developmental difficulties, multi-criteria optimization modeling of sustainability goals is essential. This chapter applies a multicriteria optimization model in the context of Nigeria, economic policy and the SDGs, incorporating economic growth, power use, GHG emissions, and employment by 2030. The results have been examined, and the model has offered mathematical support for adjustments and changes in the energy sector. Nigeria will value the development and use of renewable energy during the coming few years.

KEYWORDS

- **COVID-19**
- **multicriteria optimization modeling**
- **SDGs**
- **socioeconomic policies**
- **sustainable and economic development**

REFERENCES

Ali, I., Modibbo, U. M., Chauhan, J., and Meraj, M. (2021). An integrated multi-objective optimization modelling for sustainable development goals of India. *Environment, Development and Sustainability 23*(3), 3811–3831.

Jayaraman, R., Colapinto, C., Liuzzi, D., and La Torre, D. (2017). Planning sustainable development through a scenario-based stochastic goal programming model. *Operational Research 17*(3), 789–805.

Kamepalli, L. B., and Pattanayak, S. K. (2015). From millennium to sustainable development goals and need for institutional restructuring. *Current Science 108*(6), 1043–1044.

Modibbo, U. M., Ali, I., and Ahmed, A. (2021a). Multi-objective optimization modelling for analyzing sustainable development goals of Nigeria: Agenda 2030. *Environment, Development and Sustainability 23*(6), 9529–9563.

Modibbo, U. M., Raghav, Y. S., Hassan, M., and Mijinyawa, M. (2021b, April). A critical review on the applications of optimization techniques in the UN Sustainable Development Goals. In: *2021 2nd International Conference on Intelligent Engineering and Management (ICIEM)*. IEEE, pp. 572–576.

Modibbo, U. M., Ali, I., and Ahmed, A. (2021c). Multi-objective optimization modelling for analysing sustainable development goals of Nigeria: Agenda 2030. *Environment, Development and Sustainability*, *23*(6), 9529–9563.

National Bereau of Statistics. (2022). Retrieved from https://www.nigerianstat.gov.ng/

CHAPTER 2

SIMULATING THE IMPACTS OF LAND USE TRANSFORMATIONS ON THE FLOOD POTENTIAL USING A DISTRIBUTED HYDROLOGICAL MODEL, IN THE PALANI HILLS, SOUTH INDIA

HARIS HASAN KHAN[1], ARINA KHAN[2], ASWATHY N. S.[3], and ASIF IQUBAL[1]

[1]*Interdisciplinary Department of Remote Sensing and GIS Applications, Aligarh Muslim University, Aligarh, Uttar Pradesh, India*

[2]*Department of Geological Sciences, George Washington University, Washington, D.C., USA*

[3]*Geological Survey of India, Western Region, Jaipur, Rajasthan, India*

ABSTRACT

Transformations in land use can lead to changes in the water and sediment yield as well as modify the flood potential in river basins. Deforestation and urbanization make the soil less porous and enhance the velocity of overland flow, leading to rapid accumulation of runoff in the river channels. This work attempts to quantify the changes using a physics-based distributed hydrological model V*flo*. Land use scenarios for three different periods were developed: 1973, 2013, and potential future land use. Each

Sustainable Development: A Geospatial and Statistical Perspective. Qazi Mazhar Ali, Rizwan Ahmad, Irfan Ali, & Prasenjit Chatterjee (Eds.)

DOI: 10.1201/9781779643247-2

land use scenario was fed into the hydrological model, and storm hydrographs were simulated using 12-hour duration, spatially uniform storm as input. Statistically, land-use changes in the Porandalar basin from the 1973 to 2013 are insignificant. However, spatial patterns of land use are clearly modified. Each sub-basin, however, indicates substantial changes in land use during the 1973–2013 period, statistically as well as spatially. Marginal changes in the peak discharge and the timing to peak discharge are observed in the hydrographs of the sub-basins as well as the Porandalar basin indicating that land use transformations in this 40-year period have not significantly impacted the flood potential, except for the Gundar sub-basin, that contains the maximum forest cover.

2.1 INTRODUCTION

Hydrological processes are intricately connected to the land use and land cover of a region (Forsyth and Walker, 2008; FAO, 2008; Chang, 2006; Bruijnzeel, 2004; Reynolds and Thompson, 1988). Forests prevent soil erosion by the action of tree roots binding the soil particles and enhancing soil cohesion. The presence of forests also enhances the soil infiltration by the presence of a thick humus layer over the soil and also by the burrowing action of forest fauna and decomposing tree roots. Enhanced soil infiltration results in reduced overland flow and enhanced groundwater recharge, thereby decreasing the flood peaks and providing a prolonged and sustained source of water to the streams and rivers. Deforestation and urbanization tend to reduce the soil permeability and infiltration leading to enhanced runoff generation and reduced groundwater recharge and baseflow to streams and rivers. In the long term, this leads to severe flooding and unsustainable water resources in the river basins. Flooding as well as unsustainable water resources are major hindrances in the sustainable development of a region. The various basin management projects all over the world aim to reverse these adverse effects of urbanization and deforestation.

Hydrological simulations using appropriate models are important tools to study the impact of land use and climatic changes on the hydrological regime of rivers, as well as changes in flooding patterns. In a study conducted in a watershed in West Java, Siswanto and Frances (2019), using a distributed hydrological model TETIS, have reported a 15–40% rise in catchment water yield and a twofold increase in sediment yield, in response to urbanization and deforestation from 1994 to 2029. Khan et al. (2016) have simulated the river flow in the Shingla River Basin in

Assam, India, using the V*flo* model, mapping the flood inundation and flood hydrograph in response to storms of different intensities.

In the Porandalar catchment, the upstream parts are covered by the Shola forest and grassland biome, which is a unique biodiversity hotspot. Arasumani et al. (2018) have reported a reduction in the Shola grassland and forests due to the expansion of agriculture, timber plantations, and urbanization from 1973 to 2014. Prakasam (2010) has analyzed land use changes in the Kodaikanal taluk from 1969 to 2008 and reported a substantial loss of forests in the region at the expense of urbanization and agricultural expansion. Rangan et al. (2010) have reported substantial water resource alterations in the region due to land use changes, especially the introduction of exotic plant species since the 1960s.

The present study was designed with the following objectives:

1. Assessment of land use changes in the Porandalar basin over a period of 40 years
2. Analysis of the response of stream discharge to hypothetical rainfall events in the Porandalar basin using a distributed, finite element, physically based hydrological model—V*flo.*
3. Modeling the impact of land use transition on storm hydrograph in the Porandalar basin.

The research is conducted in the Porandalar basin (Figure 2.1), lying on the northern slopes of the eastern Palani Hills in south India. These hills form the eastward spur of the Western Ghats on the Western side of South India. The basin covers an area of 148 km^2 and is elongated in the North-South direction. Administratively, Porandalar basin lies in Palani and Kodaikanal taluks of Dindigul district.

V*flo* generates hydrographs at specific locations and provides distributed runoff maps that encompass the whole basin. The input files were initially created using SAGA GIS-System for Automated Geoscientific Analysis (Conrad et al., 2015) and subsequently loaded into V*flo* to populate the model. The overall methodology used in this work is illustrated in Figure 2.2.

The basin has a high relief (Figure 2.3a), having elevations ranging from around 500 meters to about 2500 meters above mean sea level. The upper part of the basin forms a moderate relief plateau, while the middle and lower parts have high relief and steep slopes (Figure 2.3c). The Porandalar River originates from a waterfall of 76-m height across the Gundar River. The Porandalar River enters the plains near Poolavadi

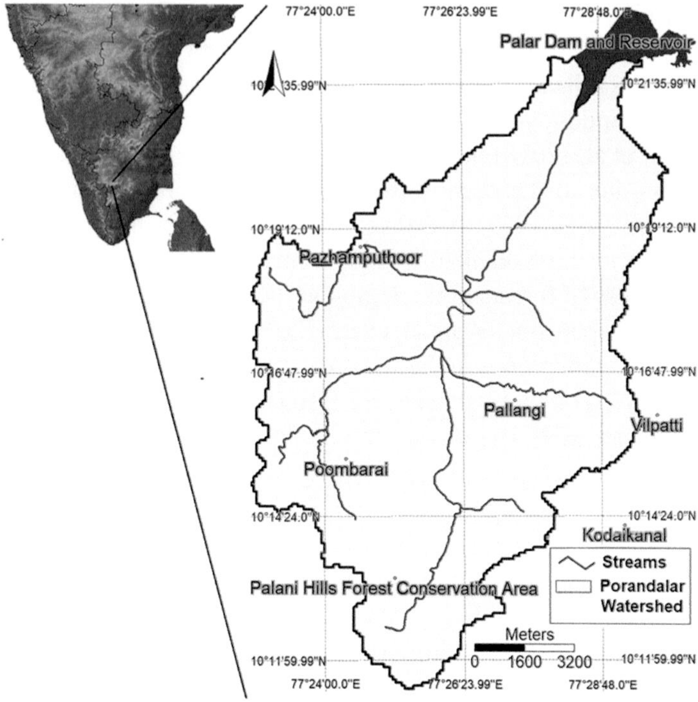

FIGURE 2.1 Porandalar basin.

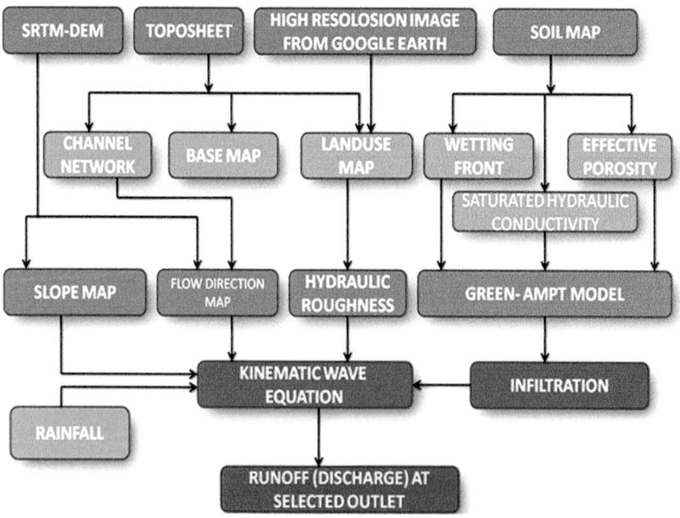

FIGURE 2.2 Flowchart highlighting the methods followed in this work.

village in Palani Taluk. The river has been dammed by the Palar Dam and forms the Palar reservoir (Figure 2.1). Poombarai, Pallangi, and Pazhamputhoor are some of the villages situated in the Porandalar basin. Kodaikanal is located on a plateau above the southern escarpment of the upper Palani Hills at an elevation of 2133 meters. The town is situated to the east of the basin. The basin exhibits a diverse and abundant collection of plant life. The woodlands are primarily characterized by the prevalence of large trees, specifically cypress, eucalyptus, and acacia. Much of the basin is also covered with scrub and grasses, forming the shola grasslands, an ecologically significant region. The basin has a monsoon-influenced subtropical highland climate. The basin has rich loamy soils in the upper part especially under the Shola forests, while in the downstream parts, the soils are dominantly clayey (Figure 2.3b).

2.2 DATA AND METHODS

This study is based on the simulation of surface runoff generation in response to input rainfall using a distributed, finite element, physically based hydrological model—V*flo*. Distributed hydrological models discretize the catchment into square grid cells arranged in a raster array. This structure is compatible with the structure of the commonly available geospatial datasets that can be easily integrated into the model. The issue of populating the model with input parameters in each grid cell is overcome by using relevant grids reclassified from thematic maps of soil, land use, and digital topography. The input datasets, with sources, used in the study are indicated in Table 2.1.

2.2.1 PRECONDITIONING OF THE DEM

Initially, the catchment boundary and the channel network were extracted from the Shuttle Radar Topography Mission (SRTM) digital elevation model. In general, channel networks and catchment boundaries extracted automatically are not in good compliance with the reality. Hence, the stream network traced from the toposheets was burned into the DEM by lowering the cell's value to the neighbor minimum value minus epsilon (set as 1) in SAGA GIS. This sets the actual channels into the correct position in the landscape, as illustrated in Figure 2.4. This was followed by

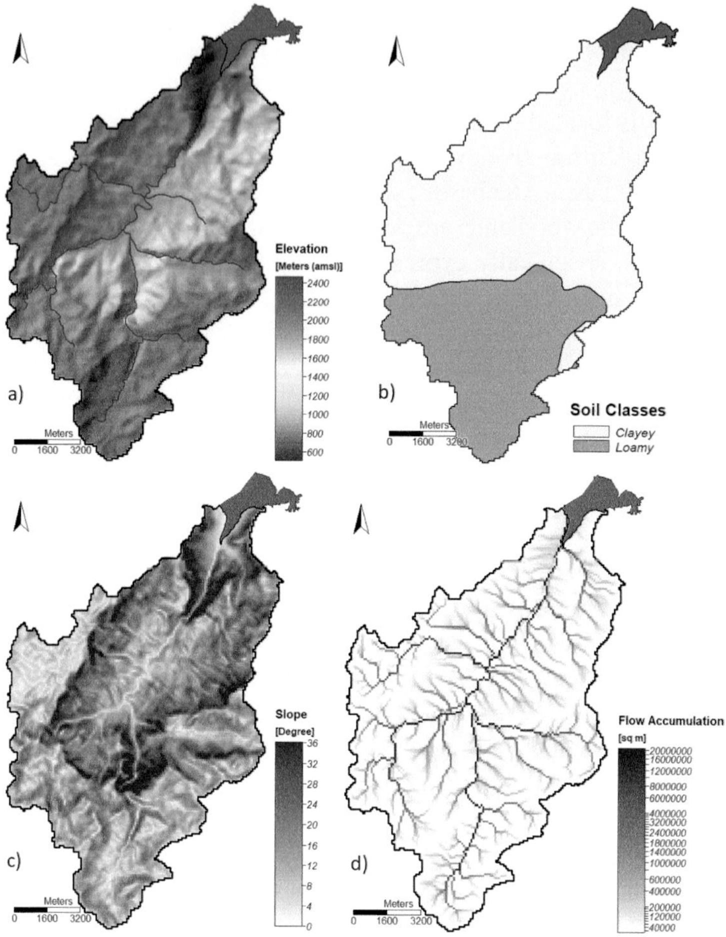

FIGURE 2.3 (a) SRTM digital elevation model, (b) soil map, (c) slope model, and (d) flow accumulation map of the study area.

TABLE 2.1 Datasets Used in the Present Study.

Dataset	Source
Baseline Spatial Information	SOI Toposheet: 58F/7 and 58F/8 (surveyed: 1972–1973)
Digital Elevation Model (DEM) (30 m)	Shuttle Radar Topography Mission
Land Use and Land Cover (LULC)	Google Earth Imagery
Soil Map (NBSS 46-b, 1:500,000)	European Digital Archive of Soil Maps, EUDASM, (http://eusoils.jrc.it/)

filling the sinks in the model using the algorithm of Planchon and Darboux (2001) to produce a downstream continuity along the channel network. Slope and Flow Accumulation were derived from the processed elevation model (Figure 2.3d). The conditioned elevation model was fed into the *Vflo* model for further processing.

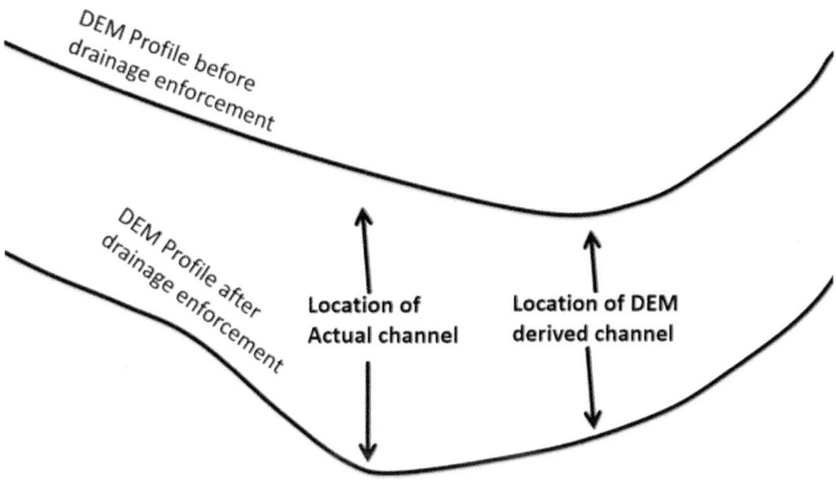

FIGURE 2.4 Process of stream burning into the digital elevation model.

2.2.2 *RAINFALL-RUNOFF MODELING*

Classically, hydrologic models have been lumped or, at best, semi-distributed, using some kind of discretization, for example, Hydrologic Response Units (HRUs). Distributed hydrologic models discretize the model domain into grid cells that are compatible with the remotely sensed data. In the United States, the prevailing method used is the unit hydrograph approach, which was initially suggested by Sherman (1932). This technique operates on the assumption that there is a linear reaction to a unit input of excess rainfall.

Distributed hydrological models refrain from using parameter/input averaging to accurately depict basin features. The computational approach utilizes the Saint Venant equations to direct the flow of runoff from one grid cell to another (as seen in Figure 2.5(left)). These equations encompass a range of models, from fully dynamic waves to simplified versions like diffusive and kinematic waves. As shown in Figure 2.5(right), each grid cell has a certain amount of water input from the rainfall, as defined by the spatially and temporally variable precipitation field. All grid cells that do not represent the basin boundary also receive some water as run-on from the upstream cells. These contributions from the rainfall and run-on are partitioned to infiltration and flow to the downstream cells. Infiltration losses are determined by the Green–Ampt infiltration model. The velocity of overland flow is governed by the manning roughness which is determined by the surface cover type.

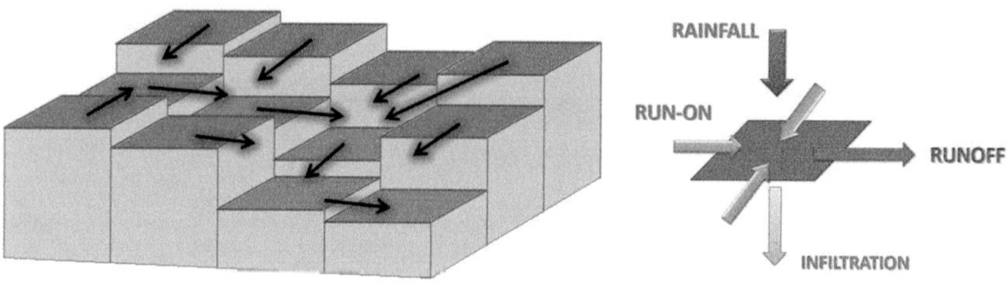

FIGURE 2.5 Runoff routing along the cells in the V*flo* model domain (left) and water balance at the grid cell scale in the V*flo* model (right).

2.2.2.1 VFLO™ MODEL FORMULATION

V*flo* employs the kinematic wave analogy (KWA), which is suitable for regions where the primary gradient is the slope of the land surface and when backwater effects are negligible. The KWA comprises the simplified momentum equation and the continuity equation. The equation that represents the continuity of overland flow resulting from excess rainfall is a one-dimensional equation:

$$\frac{\partial h}{\partial t} + \frac{\partial (uh)}{\partial x} = R - 1 \tag{2.1}$$

where R is the rainfall rate, I is the infiltration rate, h is the flow depth, and u is the overland flow velocity.

In the KWA, the relationship between the bed slope and the friction gradient is established, indicating the assumption of uniform flow. By utilizing this information in conjunction with a suitable correlation between velocity, u, and flow depth, h, such as the Manning equation, we derive

$$u = \frac{S_o^{1/2}}{n} h^{2/3} \tag{2.2}$$

where S_o is the bed slope or principal land surface slope and n is the hydraulic roughness.

Hydraulic roughness refers to the resistance to flow caused by frictional drag on the soil surface, standing vegetation, agricultural residue, pebbles, raindrop impact, and other variables in upland locations. The hydraulic roughness coefficients resulting from each of these elements contribute to the overall hydraulic resistance. The hydraulic roughness governs the speed at which surface runoff flows. Within V*flo*, a hydraulic roughness value is assigned to each land use/land cover class, utilizing the suggested Manning's coefficients (as shown in Table 2.3). The Manning equation is extensively employed to depict flow properties. The Manning hydraulic roughness coefficient, denoted as *n*, is expressed in metric units as follows:

$$n = \frac{R^{2/3} S^{1/2}}{V} \tag{2.3}$$

where S is the average slope, V is the flow velocity averaged with depth, and R is the hydraulic radius.

The velocity (u) and flow depth (h) are influenced by the topographic slope and the friction caused by the hydraulic roughness. Therefore, the KWA assumes that:

1. The slope of the water surface and the gradient of friction are aligned with the slope of the land surface.
2. The flow exhibits uniformity.
3. Access to the backwater is prohibited.

In order to relate flow depth to the topographic slope, hydraulic roughness, rainfall, and infiltration rates, Equation 2.1 is substituted into Equation 2.2 as follows:

$$\frac{\partial h}{\partial t} + \frac{S_o^{1/2}}{n} \frac{\partial h^{5/3}}{\partial x} = R - I \tag{2.4}$$

Equation 2.4 is relevant to overland flow when expressed in terms of the h variable. The gradient and hydraulic resistance exhibit spatial variability. Rainfall, infiltration, and flow depth exhibit regional and temporal variability. The finite element approach is used to describe overland flow using Equations 2.1 and 2.2, and channels are modeled using Equation 2.4 together with an appropriate form of the Manning uniform flow relation in Equation 2.2. The computation and routing of rainfall surplus through a network formulation is based on the finite element method (FEM) computational method, as defined by Vieux (1990). This process utilizes digital maps of soils, land use, terrain, and rainfall rates. Vieux (2004) discusses the creation of a rainfall-runoff model using a drainage network made up of finite components. V*flo*™ integrates the processes of scattered runoff generation and the movement of unstable flow across the channel and overland features.

2.2.2.2 INFILTRATION SUBMODEL

In order to calculate rainfall-runoff, it is important to possess a submodel of infiltration that interacts with the input of rainfall. Most hydrologic models currently employ a variation of the Green–Ampt equation to divide rainfall into runoff and infiltration components. V*flo* utilizes the Hortonian runoff concept, which states that when the rate of rainfall exceeds the rate of infiltration, runoff occurs. Various techniques for determining infiltration rates span from the comprehensive partial differential equation, commonly referred to as Richards equation, to the comparatively simplified Green and Ampt equation. The Green–Ampt equation disregards the diffusion of soil moisture over various saturation levels and only takes into account a sudden soaking front.

The Green–Ampt equation, in its rate version, is used to calculate the infiltration rate in the one-stage scenario where there are initially ponded conditions and a shallow ponded water depth:

$$f(t) = K_e \left[1 + \frac{\Psi_f \Delta\theta}{F(t)} \right] \tag{2.5}$$

where

$f(t) = dF(t)/dt$ = infiltration rate (L/T);

K_e is the effective saturated conductivity (L/T), usually taken as ½ of predicted by soil properties or measured in the laboratory accounting for the reduction due to entrapped air;

Ψ is the average capillary potential (wetting front soil suction head) (L);

$\Delta\theta$ is the moisture deficit (L/L);

$F(t)$ is the cumulative infiltration depth (L).

The Green–Ampt infiltration parameters are determined by analyzing the soil attributes using the soil map (Figure 2.3b) and Table 2.2. The soil map provides estimates for three model parameters: effective porosity (θ_e), wetting front suction head (Ψ_f), and saturated hydraulic conductivity (K_{sat}).

In the Porandalar basin, the broad soil types are loamy and clayey soils. Loamy soils are found in the upstream parts of the basin. These soils are moderately shallow, well drained, and occur on gently sloping hills. These are slightly eroded and are found associated with moderately deep, well-drained clayey soil on moderately steeply sloping hills with moderate erosion. Clayey soils are found in the central and the downstream regions of the basin. These are deep, moderately well-drained soils occurring on hill slopes. These are slightly eroded and associated with very deep, moderately well-drained calcareous loamy soils. Calcareous clay soils are found in the downstream part of the basin, and these are very deep, imperfectly calcareous, cracking clay soil found on nearly level lowlands with slightly eroded plains.

TABLE 2.2 Limits for Green and Ampt Infiltration Parameters

Soil Type	θ_e	Ψ_f (cm)	K_{sat} (cm/h)
Loamy soil	0.486 (0.334–0.534)	8.89 (1.33–59.4)	0.34 (0.2–2.5)
Clay soil	0.3 (0.269–0.501)	31.63 (6.39–156)	0.03 (0–0.10)

2.3.1 LAND USE MAPPING

Land use map mapping was conducted for 1973 and 2013. For 1973, the land use information was collected from the Survey of India topographic map numbers 58 F/7 and 58 F/8. For the land use map of 2013, high-resolution satellite imagery available in Google Earth was used. The 1973 land use polygons were digitized using the georeferenced toposheets in SAGA GIS. The 2013 land use was digitized within Google Earth and the polygons were exported as kml files, which were later cleaned and merged in SAGA GIS. Each land use class is assigned a value of Manning's coefficient based on empirical tables (Engman, 1986). The land use classes and the corresponding manning coefficient values are estimated and shown in Table 2.3.

TABLE 2.3 Land Use Classes and the Recommended Manning's Coefficient for Overland Flow

Land Use Class	Manning's Coefficient (n)
Bare soil	0.010
Cropland	0.035
Forest	0.100
Settlements	0.015
Sheet rocks	0.010
Scrub and grass	0.450
Waterbody	0.020

2.3 RESULT AND DISCUSSIONS

The land use changes have been assessed in the Porandalar basin as well as three sub-basins lying in the upstream regions of the basin. These sub-basins are Pallangi (in the east), Gundar (in the south), and Poombarai (in the west). Storm hydrographs for the outlets of the Porandalar basin as well as the sub-basins were generated in response to a synthetic storm of 12-hour duration peaking at about the eighth hour. The peak rain rate approached 1 cm/h. The impact of land use changes on hydrographs have been discussed in subsequent paragraphs.

2.3.1 LAND USE TRANSFORMATIONS

From the land use maps (Figure 2.6), it is clear that a substantial change has taken place in the land-use pattern of the region during this 40-year period. In general, the patchiness of the land use has increased during this period. This could partly be due to the difference in the source datasets that were used to generate the land-use maps. Land-use map generated from the topographic map is more uniform as compared to the land-use map derived from the high-resolution imagery in Google Earth. However, this increase in patchiness can also be attributed to enhanced human activity like plantations, settlements, etc.

Looking at the statistics of land use changes in the whole basin (Figure 2.7), however, gives another picture. From the statistics, it appears that the overall land use in the Porandalar basin has remained, more or less, stable. However, there have been major changes in the land use patterns as visible in the maps.

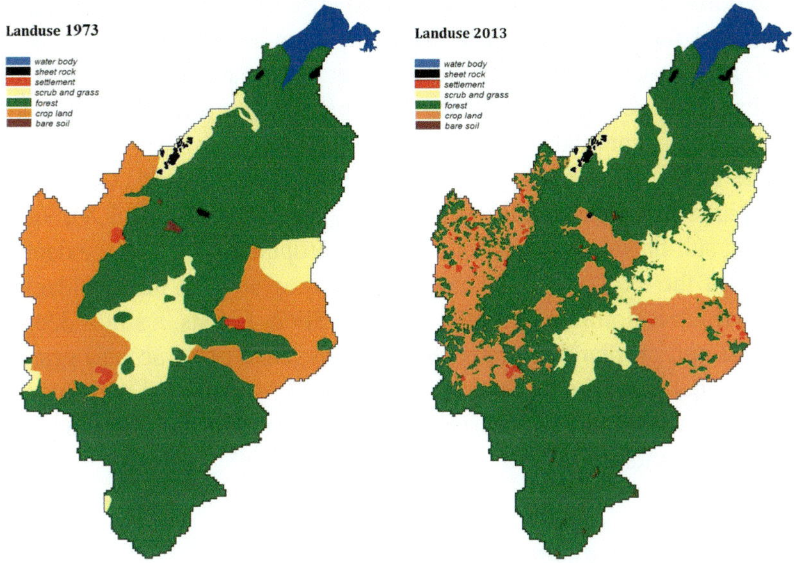

FIGURE 2.6 Land use in the Porandalar basin during the 1973 and 2013 period.

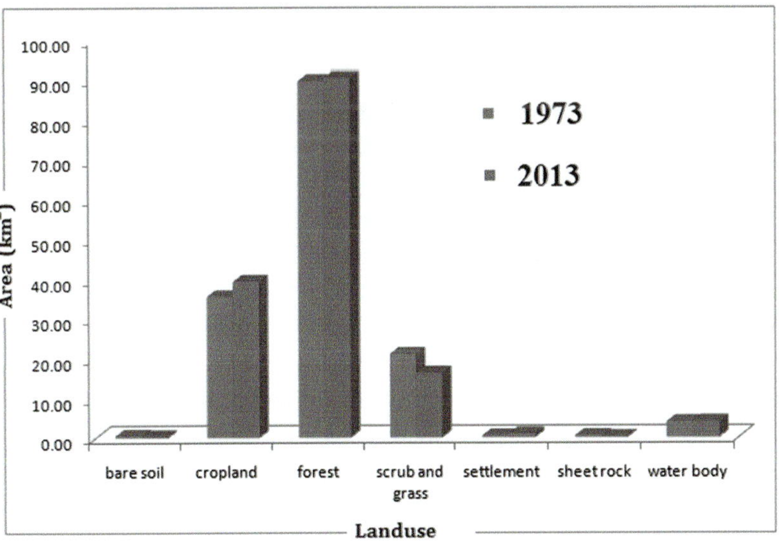

FIGURE 2.7 Land use change statistics for the Porandalar basin from 1973 to 2013.

Broadly, the forests have increased very slightly in area. In our classification, plantations have been classified within forests. Hence, the actual increase depicted by forests is attributed to the expansion of plantations. Plantations have mainly grown at the expense of cropland. Some patches of original forests have been replaced by cropland as well as scrub and grass. The Shola forests in the south appear to be marginally altered. Cropland has increased at the expense of scrub/grass and forests. Water bodies coverage has remained stable during this period.

The land use changes in the three sub-basins: Pallangi, Gundar, and Poombarai have drastically different trajectories (Figure 2.8). In the Pallangi sub-basin, which is dominated by cropland, forests have declined in area at the expense of scrub/grass and cropland. In the Gundar sub-basin, which is dominantly covered by the Shola forests, a slight decrease in forest and cropland is observed at the expense of scrub and grass. In the Poombarai sub-basin, which is dominated by cropland and forests, a reduction in cropland is observed with a concomitant increase in plantations/forests and scrub/grass.

2.3.2 HYDROGRAPH ANALYSIS

The storm hydrograph (Figure 2.9) has been simulated at the Porandalar basin outlet as well as the outlet of three sub-basins, that is, Pallangi, Gundar, and Poombarai. The hydrographs have been simulated for a 72-hour period following a storm of 12-hour duration. The rain rate increases gradually from 0.2 cm/h at the start of the simulation and approaches a maximum of 1 cm/h. The peak rain rate is achieved at approximately the 8th hour from the beginning of the simulation. The storm is set to be spatially uniform throughout the basin. The model was run three times for the simulation of three different land use conditions: 1973 land use, 2013 land use, and a probable future land use scenario with extreme deforestation and urbanization. Each simulation differed only in the manning's roughness values derived from the specific land-use conditions; other parameters and variables were similar across all three simulation runs.

The hydrograph for the 2013 land-use conditions is considered for the inter-comparison among the Porandalar basin and the sub-basins (Table 2.4). Two hydrograph parameters are selected: the peak discharges (Q_p) and the time to peak (t_p). The peak discharges are given in cubic meters per second (cumecs), while the time to peak is given in hours. The time to peak discharge is the time difference between the arrival of the peak discharge and the time of peak rainfall. Peak discharge as well as the time to peak depends upon the shape, size, drainage parameters, relief parameters, etc. of the basin as well as the soil and land-use conditions.

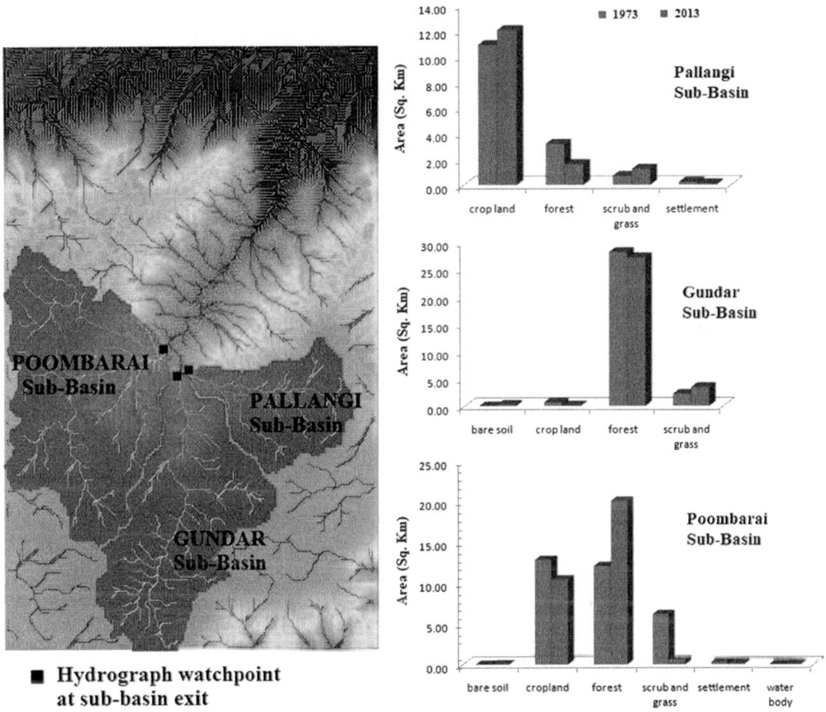

FIGURE 2.8 Land use changes in the sub-basins.

TABLE 2.4 Peak Discharge (Q_p) in Cumecs, and Time to Peak Discharge (t_p) in Hours for the Porandalar Basin and Each Sub-Basin.

	Porandalar		**Pallangi**		**Poombarai**		**Gundar**	
	Q_p	t_p	Q_p	t_p	Q_p	t_p	Q_p	t_p
1973	130.684	01:20	17.625	00:50	22.788	00:45	4.164	02:00
2013	125.817	01:20	17.627	00:50	22.92	00:45	4.14	02:00
Future	158.846	01:00	19.153	00:45	27.982	00:35	4.603	01:15

The hydrograph (Figure 2.9) depicts a rise in stream discharge from almost zero to very high values. The discharge in all basins is observed to increase very rapidly only after the arrival of peak rainfall. The zero discharge is due to the nonassignment of the baseflow component in the model. The discharge in the main hydrograph returns to zero at approximately the 42nd hour. This is an unrealistic scenario, which is observed, again, due to the nonassignment of the baseflow.

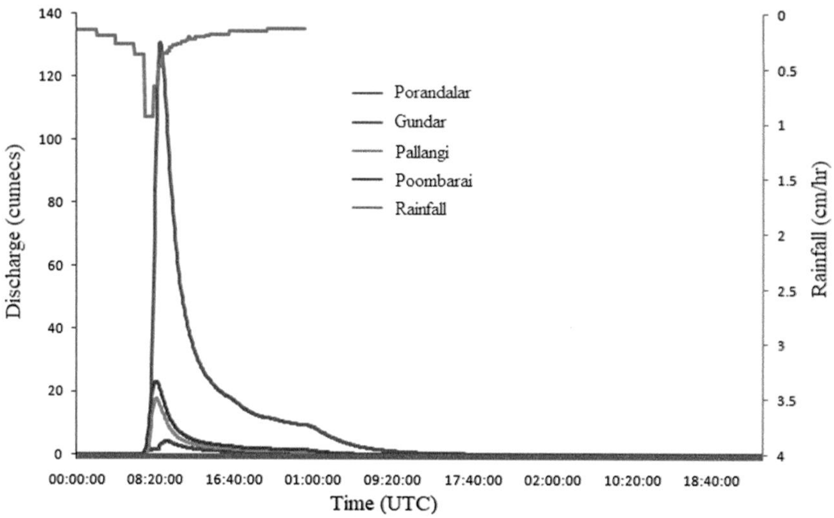

FIGURE 2.9 Hydrograph comparison for outlets of each sub-basin and the outlet of the Porandalar basin.

In general, hydrograph peaks are increasing with increasing area of the basins. However, the Gundar sub-basin, which is larger than the Pallangi sub-basin, has a subdued peak of 4.164 cumecs. This can be attributed to the elongated nature of the Gundar sub-basin as well as the dominance of forests and presence of loamy soils. In elongated catchments, the surface runoff approaches the outlet in a sequential manner, leading to less prevalence of flooding as well as a subdued and prolonged peak flow. The forested landscape has a thick layer of humus as well as a very porous topsoil due to the burrowing activity of animals and decomposition of tree roots. This results in high infiltration rates and less or almost no overland flow generation, leading to low-peak runoff at the catchment outlet.

The Pallangi sub-basin has the maximum peak discharge, despite its small size, due to the dominance of cropland. Cropped land offers relatively low resistance to overland flow and has a relatively low manning's roughness as compared to forests and scrub/grass. Moreover, the presence of clayey soils also leads to less infiltration and enhanced runoff production.

The peak discharge for the Porandalar basin is far greater than the combined peaks of the three sub-basins. This indicates an enhanced runoff production in the downstream regions of the basin. This is possibly due to the prevalence of clayey soils in the downstream parts of the basin.

The time to the arrival of peak discharge also varies within the sub-basins. The Poombarai sub-basin has the shortest time to peak of 45 min, followed by the Pallangi, with flood peak arriving after 50 min of the peak rainfall. As the peak discharge travels downstream, it quickly arrives at the outlet of the Porandalar basin, with a time to peak of 70 min. However, the most noticeable is the maximum time to peak (120 min) of the Gundar sub-basin that can be attributed to elongated shape of the sub-basin as well as the presence of forests that reduce the overland flow velocity.

2.3.3 *IMPACT OF LAND-USE CHANGES*

The impact of land-use transformations on storm hydrographs has been studied throughout the world, and the general consensus is that deforestation and urbanization both increase the flood peak and reduce the travel time of the flood wave. In the Porandalar basin (Table 2.4), the peak discharge is seen to be reduced by about 5 cumecs, while time to peak is unaffected in the 40-year period from 1973 to 2013 in response to land-use changes, which are, in effect, reorganization of the land-use parcels (Figure 2.10). Since the land-use statistics have not changed much, a similar response is seen in the hydrograph. Similarly, in the Pallangi, Poombarai, and Gundar sub-basins, the time to peak is unaffected, while the peak discharge is very marginally modified.

Due to rapid population growth, a potential land-use scenario expected in future is the expansion of croplands and urban centers. The urban centers could potentially develop around present villages, while cropland can potentially expand into the grasslands and forests. Such a land-use scenario was developed by making buffers around the present settlements and assigning all the forests and grasslands as croplands. This potential future land use map for the Porandalar basin was used to assign new manning's roughness values to the grid cells of the model. The results (Table 2.4 and Figure 2.10) indicate that the peak discharge increases by approximately 33 cumecs, with respect to the 2013-peak discharge. The time lag in the arrival of peak discharge decreases by 20 min. An increase in 33 cumecs of river discharge needs to be adjusted in the Palar reservoir at the outlet of the basin. These values are not very significant considering the magnitude of the land-use changes imposed on the model.

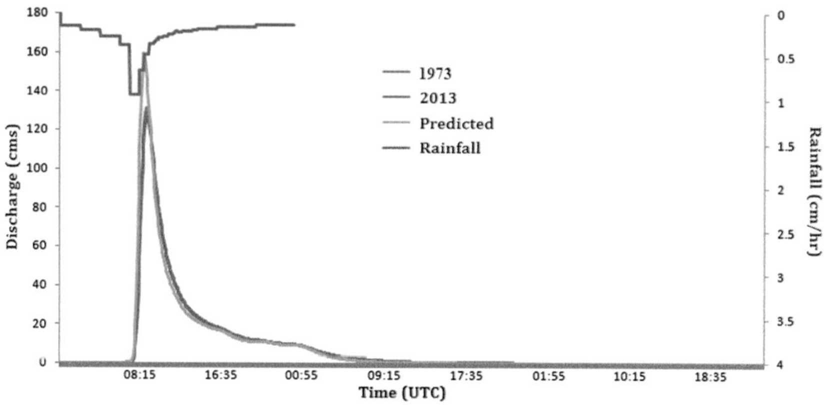

FIGURE 2.10 Hydrograph for the Porandalar basin.

In the Pallangi sub-basin (Figure 2.11), the predictions for the potential future land use indicate that the peak discharge has marginally increased from 17.63 to 19.15 cumecs, and the time of arrival of the peak discharge has reduced by 5 min. In the Poombarai sub-basin (Figure 2.12), the peak discharge increases very marginally from 22.79 cumecs in 2013 to 22.92 cumecs for the future land-use scenario, while the time to peak discharge is reduced by 10 min. In the Gundar sub-basin (Figure 2.13), the peak discharge increases from 1.14 cumecs for 2013 land use to 1.60 cumecs for the future land-use scenario. The time to the peak discharge is reduced by almost 45 min, which is a significant increase. Among the sub-basins, it can be observed that the most prominent changes in hydrographs of 2013 and the future land-use scenario are observed in the Gundar sub-basin. This is attributed to the dominant presence of forests in the Gundar sub-basins as compared to the other two sub-basins. A simulation with these forests turning into cropland would have the greatest effect on the sub-basin with the largest forest cover. Hence, both the peak discharge and the time to peak are drastically modified for the future land-use scenario in the Gundar sub-basin.

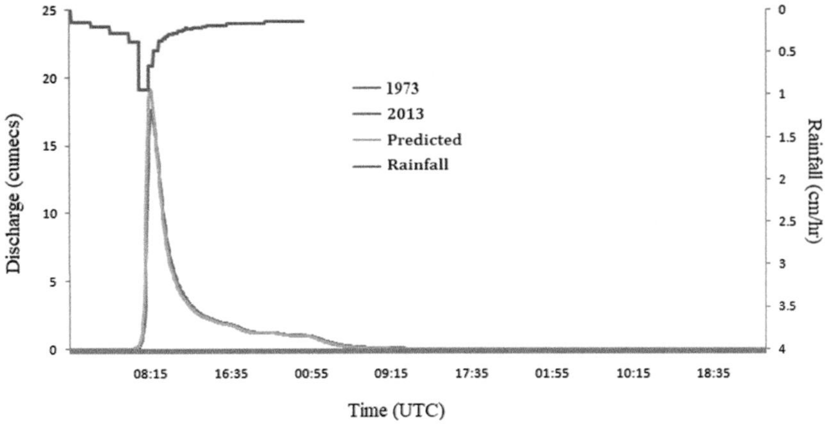

FIGURE 2.11 Hydrograph for the Pallangi sub-basin.

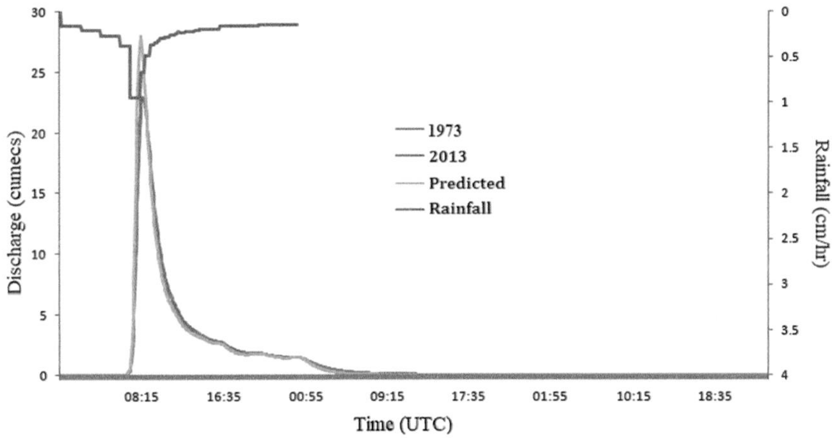

FIGURE 2.12 Hydrograph for the Poombarai sub-basin.

Storm hydrographs develop in response to single precipitation events, which are usually short-lived. When these storms are intense, they tend to produce flash floods, which is represented by a very sudden and sharp rise in the hydrograph. In low-relief basins, with thick forest cover and porous soils, sharp-crested hydrographs are not usually observed due to the subduing effect of the low-relief, porous soil, and thick vegetation. High-relief basins, on the other hand, tend to produce flash floods that could be worse if the basin is nonforested. In the case of the Porandalar basin and its sub-basins, it can be clearly observed that even after imposing intense land-use transformations, the model results do not indicate substantial changes in the hydrographs. This is an indication that the high-relief and rugged topography is playing a dominant role in shaping the storm hydrograph, while land use plays a secondary role in the Porandalar basin. Rugged basins usually have thin soils due to frequent mass movements and soil loss, which leads to less infiltration and greater surface runoff. Also, steep slopes tend to push the overland flow quickly over the hillslopes to nearby channels and lead to water piling up in the channels and creating a flood wave.

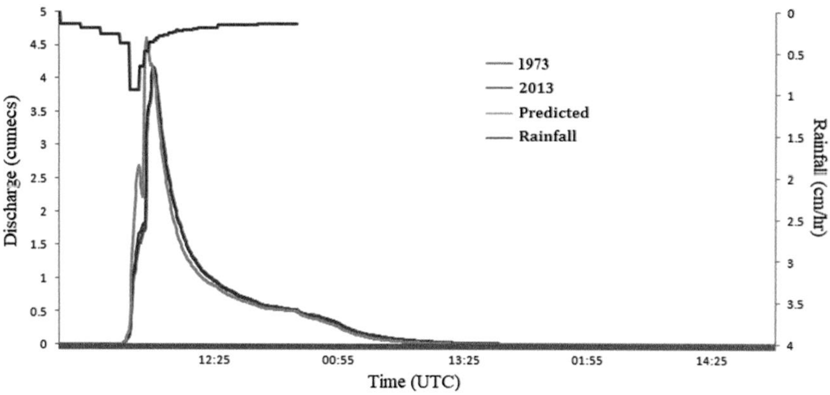

FIGURE 2.13 Hydrograph for the Gundar sub-basin.

2.4 CONCLUSIONS

The land-use changes in the Porandalar basin have been assessed for the time period 1973–2013. It is observed that the land-use pattern has considerably changed in the basin, but the area of the major land-use categories has remained almost the same for this period. On analyzing the land use within different sub-basins, it was found that the aerial coverage has changed significantly. This leads to the conclusion that the trends of land use changes in different sub-basins are mutually inverse, so that, statistically, the overall land use change in the Porandalar basin is insignificant.

The stream hydrographs for the Porandalar basin as well as the sub-basins simulated for the 1973 and 2013 land-use scenarios show very little change. Only a marginal change in the peak discharge and the time of arrival of peak discharge is observed from the 1973 scenario to the 2013 scenario, even in sub-basins where the land-use changes are significant. The peak discharge for a probable future scenario, where all the forests have been cleared, is also not in phase with the magnitude of the land-use change imposed on the system except for the Gundar sub-basin which is dominated by the shola forests. All these observations suggest that land use plays a secondary role in characterizing the stream hydrograph in this mountain basin. It appears that the response of the stream hydrograph (peak discharge as well as the timing of peak) is more influenced by the steep topography of the basin than by the land use and its transformations.

KEYWORDS

- **land use**
- **hydrograph**
- **hydrological model**
- **flood potential**
- **Palani Hills**

REFERENCES

Arasumani, M., Khan, D., Das, A., Lockwood, I., Stewart, R., Kiran, R. A., et al. (2018). Not seeing the grass for the trees: Timber plantations and agriculture shrink tropical montane grassland by two-thirds over four decades in the Palani Hills, a Western Ghats Sky Island. *PLoS One 13*(1), e0190003. https://doi.org/10.1371/journal.pone.0190003

Bruijnzeel, L. A. (2004). Hydrological functions of tropical forests: Not seeing the soil for the trees? *Agriculture, Ecosystems & Environment 104*, 185–228.

Chang, M. (2006). *Forest Hydrology*, 2nd edn, Boca Raton: CRC Press.

Conrad, O., Bechtel, B., Bock, M., Dietrich, H., Fischer, E., Gerlitz, L., Wehberg, J., Wichmann, V., and Boehner, J. (2015). System for automated geoscientific analyses (SAGA) v. 2.1.4. *Geoscientific Model Development 8*, 1991–2007, https://doi.org/10.5194/gmd-8-1991-2015.

Engman, E. T. (1986). Roughness coefficients for routing surface runoff. *Journal of Irrigation & Drainage Engineering 112*(1), 39–53.

FAO. (2008). *Forests and Energy*. Forestry Paper 154, Rome: FAO.

Forsyth, T., and Walker, A. (2008). *Forest Guardians, Forest Destroyers: The Politics of Environmental Knowledge in Northern Thailand.* Seattle: University of Washington Press.

Khan, A. S., Khan, A. B., and Khan, H. H. (2016). Flood inundation mapping of Shingla river basin in Assam using a distributed hydrological model. *The Ecoscan 10*(3&4), 405–409.

Planchon, O., and F. Darboux. (2001). A fast, simple and versatile algorithm to fill the depressions of digital elevation models. *Catena 46*, 159–176.

Prakasam, C. (2010). Land use and land cover change detection through remote sensing approach: A case study of Kodaikanal taluk, Tamil nadu. *International Journal of Geomatics and Geosciences 1*(2), 150–158.

Rangan, H., Christian A. K, and L. A. (2010). Forest plantations, water availability, and regional climate change: Controversies surrounding Acacia mearnsii plantations in the upper Palani Hills, southern India. *Regional Environmental Change 10*, 103–117.

Reynolds, E., and Thompson, F., Eds. (1988). *Forests, Climate and Hydrology: Regional Impacts*. Tokyo: United Nations University.

Sherman, L. K. (1932). Streamflow from rainfall by the unit-graph method. *Engineering News-Record 108*, 501–505.

Siswanto, S. Y., and Francés, F. (2019). How land use/land cover changes can affect water, flooding and sedimentation in a tropical watershed: A case study using distributed modeling in the Upper Citarum watershed, Indonesia. *Environmental Earth Sciences 78*, 550

Vieux, B. E., Bralts, V. F., Segerlind, L. J., and Wallace, R. B. (1990). Finite element watershed modeling: One-dimensional elements. *Journal of Water Resources Planning and Management 116*(6), 803–819.

Vieux, B. E. (2004). *Distributed Hydrologic Modeling Using GIS*. Norwell, MA, USA: Kluwer Academic Publishers, Water Science Technical Series. ISBN 0-7923-7002-3.

CHAPTER 3

DECADAL LAND USE/LAND COVER CHANGE DETECTION USING REMOTE SENSING & GIS TECHNIQUE OF KARNATAKA, SOUTH INDIA

QAZI ZEESHAN AHMAD JALALI[1], MOHAMMED IRSHAD[2], RIZWAN AHMAD[1], P. S. ROY[2], and MUHAMMAD ZAID QAMAR[1]

[1]*Interdisciplinary Department of Remote Sensing & GIS Applications, Aligarh Muslim University, Aligarh, Uttar Pradesh, India*

[2]*International Crops Research Institute for the Semi-Arid Tropics, Patancheru, Hyderabad, India*

ABSTRACT

In numerous applications, including land planning and global warming studies, the mapping of changes in land use and land cover (LULC) on a regional scale is crucial. LULC changes can occasionally interfere with the climate and natural processes, which have an adverse impact on regional patterns. In this study, we use remote sensing and geographic information systems (GIS) to analyze these LULC changes in the Raichur district of Karnataka. In the first stage of this investigation, data from Landsat 8 collected in December 2015 are subjected to a supervised classification algorithm. With the help of ground control points retrieved from Google Earth Pro, also dated December 2015, the maximum likelihood approach is used to classify the image using the six bands of the Landsat image. The data of the shift in land use and land cover are examined in depth in the

Sustainable Development: A Geospatial and Statistical Perspective. Qazi Mazhar Ali, Rizwan Ahmad, Irfan Ali, & Prasenjit Chatterjee (Eds.)

DOI: 10.1201/9781779643247-3

study's second phase. The third step concludes with a map that shows how the land cover has changed. The complexity and dynamics of changes in the region's land use and land cover are better understood, thanks to this tripartite approach.

3.1 INTRODUCTION

Land use and land cover changes (LULC) play an important role in studying local and global changes. Environmental problems linked to human needs such as deforestation, loss of biodiversity, etc. are often linked to changes in LULC. Therefore, the information available on LULC changes in an area can provide the idea of managing and planning for the future through correct decision-making. The growing population and the growing need for water and arable land are putting pressure on the LULC. This pressure means that changes to the LULC are unforeseen and uncontrolled. The changes in LULC are mainly caused by poor management of land, urban, and forest areas, which causes most environmental problems. Remote sensing and GIS are powerful tools for obtaining precise and timely information on the space-time distribution of LULC over large areas. GIS provides a simple environment for collecting, storing, viewing, and analyzing the digital data needed to detect changes. The purpose of the change detection process is to detect LULC changes on digital images of the same area and at different times. In this study, the detection of changes was calculated using classified images from different years. The objective of the study is to analyze changes in LULC using satellite images and GIS in the Raichur district (Karnataka), which is bordered by the Krishna River to the north and the Tungabhadra River to the south. To achieve this objective, Landsat 8 data collected in December 2015 and classified data from 1985, 1995, and 2005 are used. The detection of changes in LULC is made easier with the use of maximum likelihood classification and comparative change detection approaches.

Initially, in the absence of data and remote computer support, using tracing paper and topographic sheets, modifications to land use and cover were discovered. But this method took a lot of time and effort to study the big ones. Standard soil methods for land-use mapping are also time-consuming and rarely used. With the emergence of satellite remote-sensing technologies, it is therefore relatively easier to create accurate land-use maps for land use and to monitor changes at regular time intervals. This method is the only option for exploring remote areas. With multitemporal analysis, remote sensing offers a different perspective on the development of rural areas. The Raichur district has mainly rice crops. The area is largely covered with arable land. Raichur appears as a city with the commitment of an urban landscape. The evaluation of the GIS and remote sensing of the study area shows that the city is still young and that there is a certain chance of protecting it from the harmful effects of urbanization, popularly known as urban sprawl. Urban sprawl has been criticized for removing agricultural land, affecting water quality, and causing air pollution. It is therefore important to see the changes taking place in the Raichur district and to plan for symmetrical growth in urban areas without affecting agricultural productivity. The process of identifying changes in a subject's or phenomenon's state over a range of observational periods is known as "change detection." Timely and accurate detection of changes in Earth's surface features also promotes more effective resource management and utilization because it forms the basis for a deeper comprehension of the connections and interactions between human activity and natural occurrences. Land cover modifications brought on by changes in land use may not always indicate land degradation. However, a change in land use patterns can result in changes to the land cover that have an impact on biodiversity, water, and other interconnected processes affecting the climate and biosphere. The LULC change detection study is the process that helps in determining the changes related to LULC. Remote sensing and GIS technology allow us to examine the changes of LULC. GIS provides effective methods for investigating change detection in LULC and tools for land-use planning that can be used to examine changes of land cover in less time, at low cost, and with greater precision.

3.2 STUDY AREA

Raichur city, which is part of Karnataka State's North Eastern area, is bounded by latitudes of 16°9′17.804″–16°15′2.521″ North and longitudes of 77°18′23.912″–77°24′43.895″ East (Figure 3.1). The Krishna and the Tungabhadra, two important rivers, are sandwiched between it. Rice is the main crop grown in the city's agricultural land, which is its main source of income. Around 232,000 people called the city home as per the 2011 census. The soil on the property includes a mixture of hues, including light green, reddish-brown, and black. The majority of the area is covered in black soil, especially near the hornblende schists and gneiss rocks. Averaging 382 m above sea level, the geographic gradient mainly slopes from the northwest to the southeast. The climate in Raichur is characterized by mild winters and sweltering summers. December is the coldest month, with an average daily low temperature of 17.7°C. The warmest month, on the other hand, is May, with an average daily maximum temperature of 39.8°C with frequent increases to 45.0°C. Raichur is located in the northeastern dry agroclimatic zone and is subject to drought. The city experiences 621 mm of rainfall on average per year, spread out across roughly 49 days. Approximately 67% of the yearly rainfall is the result of the southwest monsoon, which lasts from June to September. Approximately 24% of the total precipitation falls during postmonsoon or northeast monsoon (Central Ground Water Board "CGWB," 2008).

3.3 DATA AND METHODOLOGY

A flowchart of the general steps described in this work to derive statistics on the land-use model of the area is shown in Figure 3.2. For this study, two Landsat 8 satellite images (path 144, row 049) and (path 145, row 049) of the Raichur district for 2015 and 2020 were taken and classified image of the Raichur district from 1985, 1995, and 2005 were used from ORNL DAAC (Distributed Active Archive Center for Biogeochemical Dynamics) https://daac.ornl.gov/. Landsat 8 satellite imagery was downloaded from https://earthexplorer.usgs.gov/. Landsat 8 images have a spatial resolution of 30 m. Satellite images were brought to UTM projection 44 N zone (Figure 3.1).

The following three software were used in this project:

1. ERDAS IMAGINE 9.1: This was used to display and mosaic the images.
2. ArcMap 10.3: This was used to extract the area of interest (AOI), perform classification, and create LULC map.
3. Google Earth Pro: This was used to take the ground control points (GCP).

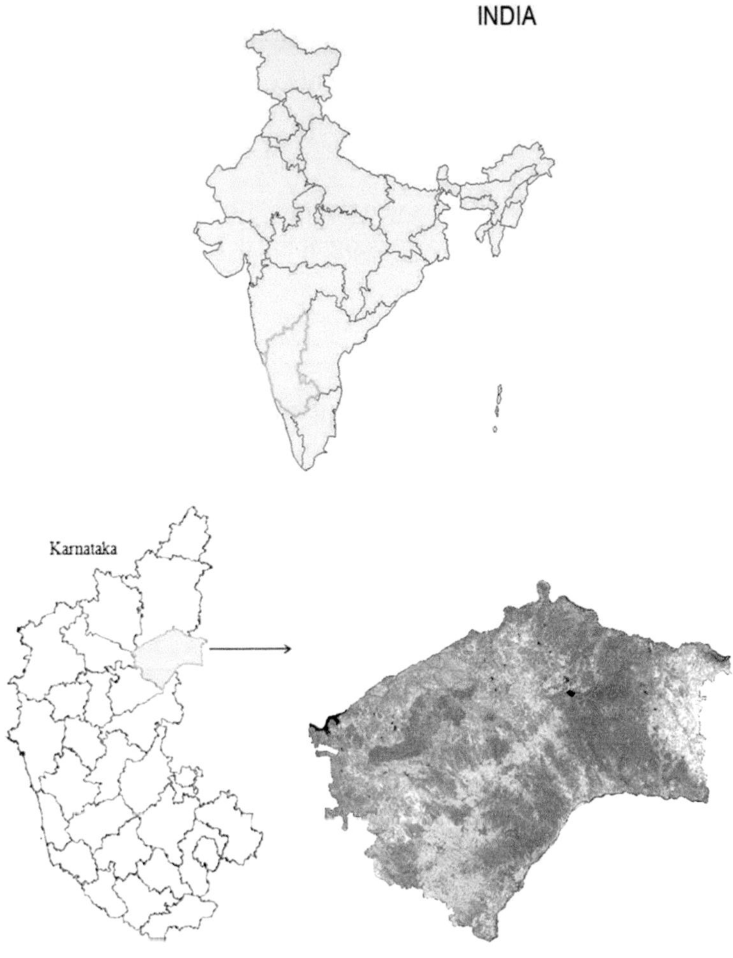

FIGURE 3.1 Location map of the study area's geographical position.

3.4 DESCRIPTION OF THE PARAMETERS

3.4.1 REMOTE SENSING

Remote sensing is the practice of gathering information about a subject, a space, or a phenomenon without coming into contact with it directly. According to Zubair and Opeyemi (2006), this term is frequently connected, especially to the measurement of interactions between the materials on the earth's surface and electromagnetic energy. Additionally, it encompasses the science of compiling, analyzing, and interpreting data.

3.4.2 GEOGRAPHIC INFORMATION SYSTEM

A computerized system called a geographic information system (GIS) is used to store, analyze, and display geographic data.

3.4.3 LAND USE

This term "land use" refers to the different ways that people use and exploit the land and its resources.

3.4.4 LAND COVER

This describes the surface of the Earth in its physical or natural state.

3.4.5 GROUND CONTROL POINTS

A GCP is a place on the surface of the Earth with a precise coordinate system and a known location. It is used to georeference picture data sources such as scanned maps and remotely sensed photos.

3.4.6 MOSAICKING

According to a tutorial on remote-sensing techniques using Arcview 3.2 and ERDAS IMAGINE 8.4 from 2001, mosaicking is the act of combining two georeferenced images into a single image (Figure 3.2).

3.5 RESULT AND DISCUSSION

In the present study, we have supervised the classification of 2015 Landsat 8 image of the Raichur district using ArcMap, ERDAS, and Google Earth Pro. We have also calculated the changes in LULC over 30 years from

1985 to 2015 (Tables 3.1 and 3.2) of the Raichur district to understand the trend of land cover in 30 years. From the different class change calculations, we have plotted various graphs and tables which show the change detection of each class of every 10, 20, and 30 years. So we will discuss each class result separately.

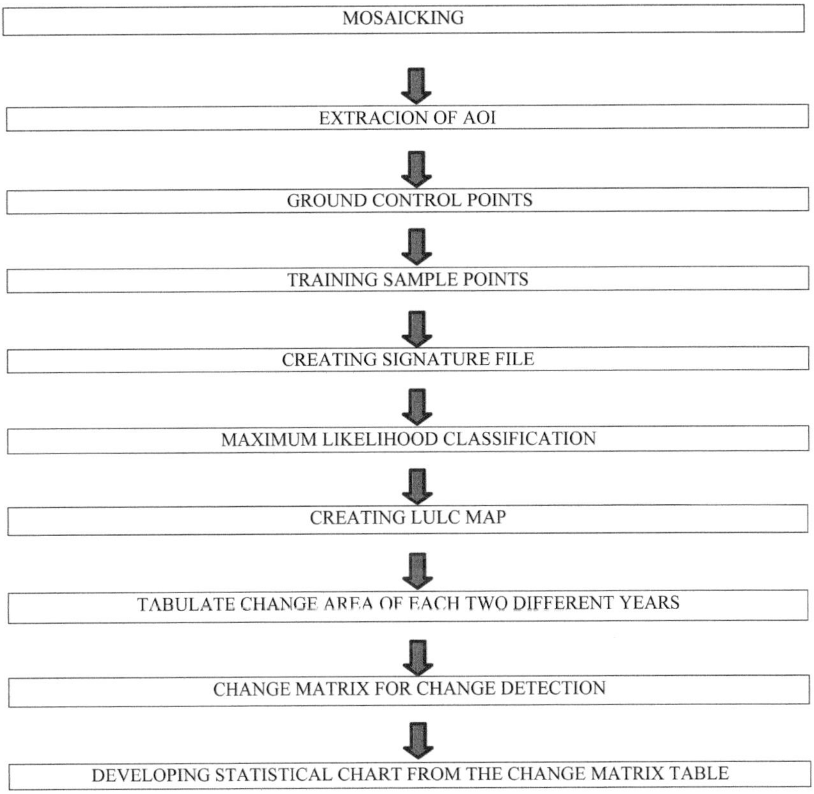

FIGURE 3.2 Flowchart showing different steps adapted to follow methodology in the study area.

Figure 3.3 shows the change in cropland from 1985 to 1995 (0.2%), from 1995 to 2005 (0.9%), and from 1985 to 2005 (1.1%). There was not much change in cropland. During the years 2005–2015 (–36.2%), 1995–2015 (–35.6%), and 1985–2015 (–35.5%), the area decreased significantly and turned into different classes. The information in the change matrix shows that the class to which the cropland area converted most was shrub-land.

TABLE 3.1 Change matrix of land use land cover (Sq.km) in 1985 to 1995

Land Class		Land Class 1995								Grand Total	Change %
		BL	**BU**	**CL**	**FL**	**MF**	**SL**	**WB**	**WL**		
Land class 1985	BL	124.82	0	0.07	0.04	0.01	0.01	1.87	0	126.82	2.4
	BU	0	**27.94**	0.06	0.01	0	0.01	0	0	28.02	30.7
	CL	0.08	8.86	**7041.47**	33.27	0.12	2.56	49.12	0.83	7136.31	0.2
	FL	4.99	1.02	96.55	**399.77**	0.02	2.32	2.91	0	507.58	-13.9
	MF	0	0	1.92	1.86	**164.76**	16.25	1.73	0	186.52	-11.4
	SL	0	0	0.01	0.18	0.31	**205.2**	0.44	0	206.14	9.5
	WB	0.04	2.64	12.09	1.48	0.02	1.44	**250.41**	0.47	268.59	12.4
	WL	0	0	0	0	0	0	0.14	**26.2**	26.34	4.2
	Grand total	129.93	40.46	7152.17	436.61	165.24	227.79	306.62	27.5	8486.32	

TABLE 3.2 Change matrix of land use land cover (Sq.km) in 1985 to 2005

Land Class		Land Class 2005									
		BL	BU	CL	FL	MF	SL	WB	WL	Grand Total	Change %
Land class 1985	BL	**122.17**	0.02	1.99	0.28	0.04	0.23	2	0.09	126.82	2.4
	BU	0.02	27.6	0.3	0.06	0	0.01	0.02	0	28.01	47.3
	CL	2.1	12.82	**7012.47**	26.18	1.27	6.9	72.03	2.12	7135.89	1.1
	FL	5.17	10.04	171.06	**312.72**	0.23	5.18	3.05	0.01	507.46	−32.8
	MF	0.03	0	2.58	0.52	**177.77**	4.05	1.56	0	186.51	−3.5
	SL	0.28	0.02	4.33	0.25	0.42	**200.1**	0.76	0	206.16	5.5
	WB	0.13	2.66	21.63	1.01	0.31	1.65	**240.2**	0.52	268.11	16.2
	WL	0.09	0	1.19	0.01	0	0	0.17	**24.93**	26.39	4.6
	Grand total	129.99	53.16	7215.55	341.03	180.04	218.12	319.79	27.67	8485.35	

TABLE 3.3 Change matrix of land use land cover (Sq.km) in 1995 to 2005

Land Class		Land Class 2005								Grand Total	Change %
		BL	**BU**	**CL**	**FL**	**MF**	**SL**	**WB**	**WL**		
Land class 1995	BL	**126.92**	0.02	2	0.38	0.04	0.23	0.25	0.09	129.93	0.1
	BU	0.02	**39.8**	0.51	0.07	0	0.01	0.05	0	40.46	23.8
	CL	2.04	4.61	**7089.26**	22.45	1.73	4.5	25.86	1.35	7151.8	0.9
	FL	0.37	8.67	104.89	**317.48**	1.76	3	0.4	0.01	436.58	−21.9
	MF	0.03	0	1.13	0.18	**163.52**	0.24	0.14	0	165.24	8.2
	SL	0.28	0.01	4.57	0.18	12.4	**209.8**	0.5	0	227.74	−4.2
	WB	0.24	0.02	12	0.28	0.6	0.4	**292.55**	0.09	306.18	4.3
	WL	0.09	0	1.2	0.01	0	0	0.06	**26.13**	27.49	0.6
	Grand total	129.99	53.13	7215.56	341.03	180.05	218.18	319.81	27.67	8485.42	

TABLE 3.4 Change matrix of land use land cover (Sq.km) in 1985 to 2015

Land Class		Land Class 2015								Grand Total	Change %
		BL	**BU**	**CL**	**FL**	**MF**	**SL**	**WB**	**WL**		
Land class 1985	BL	**25.4**	6.42	14.38	6.6	43.36	26.7	0.01	3.95	126.82	80.8
	BU	0.61	**19.87**	5.22	0.29	0.55	0.3	0.04	1.14	28.02	94.4
	CL	560.8	381.71	**4245.95**	312.88	546.56	1000.71	11.74	75.88	7136.23	−35.5
	FL	26.96	38.87	166.94	**137.21**	74.25	37.03	0.14	26.2	507.6	−5.9
	MF	9.79	1.95	11.24	5.41	**94.52**	61.34	0	2.27	186.52	77.9
	SL	15.27	7.21	28.54	9.36	72.38	**69.32**	0.1	4	206.18	82.9
	WB	21.12	36.6	126.98	3.48	11.53	14.91	**46.02**	7.89	268.53	−78.4
	WL	0.95	5.32	4.49	2.1	1.82	0.52	0	**11.19**	26.39	80.1
	Grand total	660.9	497.95	4603.74	477.33	844.97	1210.83	58.05	132.52	8486.29	

TABLE 3.5 Change matrix of land use land cover (Sq.km) in 1995 to 2015

Land Class		Land Class 2015								Grand Total	Change %
		BL	BU	CL	FL	MF	SL	WB	WL		
Land class 1995	BL	**25.78**	6.84	14.6	7.59	44.29	26.14	0.01	4.68	129.93	80.3
	BU	0.95	**26.84**	9.5	0.35	0.74	0.47	0.29	1.33	40.47	91.8
	CL	560.97	376.18	**4258.85**	321.1	553.17	998.7	7.76	75.45	7152.18	−35.6
	FL	23.2	34.18	135.23	**126.38**	61.65	31.52	0.22	24.24	436.62	8.5
	MF	8.63	1.63	9.42	4.89	**84.45**	54	0	2.22	165.24	80.4
	SL	16.54	7.92	32.11	10.73	81.38	**74.88**	0.11	4.1	227.77	81.2
	WB	23.78	38.58	139.23	4.18	17.45	24.59	**49.56**	9.2	306.57	−81.1
	WL	1.07	5.8	4.78	2.1	1.82	0.54	0.09	**11.3**	27.5	79.2
	Grand total	660.92	497.97	4603.72	477.32	844.95	1210.84	58.04	132.52	8486.28	

TABLE 3.6 Change matrix of land use land cover (Sq.km) in 2005 to 2015

	Land Class	Land Class 2015									
		BL	BU	CL	FL	MF	SL	WB	WL	Grand Total	Change %
Land class 2005	BL	**25.8**	6.84	14.64	7.62	44.26	26.14	0.01	4.68	129.99	80.3
	BU	2.38	**29.86**	12.21	1.09	4.06	1.51	0.3	1.74	53.15	89.3
	CL	561.51	382.25	**4277.56**	339.75	562.05	1004.23	7.02	81.68	7216.05	−36.2
	FL	19.45	23.18	104.4	**105.94**	47.94	23.03	0.04	17.22	341.2	28.5
	MF	9.14	1.78	10.38	5.13	**91.45**	59.93	0	2.24	180.05	78.7
	SL	16.1	7.9	31.88	11.38	75.86	**70.6**	0.11	4.32	218.15	81.9
	WB	25.47	40.36	148.04	4.32	17.52	24.87	**50.5**	9.24	320.32	−81.9
	WL	1.08	5.78	4.8	2.12	1.84	0.53	0.1	**11.41**	27.66	79.1
	Grand total	660.93	497.95	4603.91	477.35	844.98	1210.84	58.08	132.53	8486.57	

TABLE 3.7 Change matrix of land use land cover (Sq.km) in 1985 to 2020

Land Class		Land Class 2020								Grand Total	Change %
		BL	**BU**	**CL**	**FL**	**MF**	**SL**	**WB**	**WL**		
Land class 1985	BL	**95.4**	6.42	5.38	6.6	4.36	6.1	0.01	3.25	126.82	60.2
	BU	0.61	**21.87**	5.22	0.29	0.55	0.3	0.04	1.14	28.02	73.8
	CL	60.8	31.71	**6945.95**	12.88	2.56	54.71	11.74	13.88	7136.23	4.5
	FL	12.96	21.87	194.94	**137.21**	2.25	113.03	1.14	16.2	507.6	−49.7
	MF	6.79	1.95	61.24	25.41	**7.32**	41.34	0	4.27	186.52	−89.4
	SL	15.27	7.21	78.54	29.36	2.18	**69.32**	0.1	4	206.18	41.3
	WB	11.12	9.6	136.98	23.48	1.13	5.91	**41.02**	9.89	268.53	−79.8
	WL	0.95	3.32	12.49	4.1	1.22	0.52	0	**5.19**	26.39	50.2
	Grand total	203.1	106.79	7454.56	255.35	19.68	291.36	54.26	53.02	8486.29	

TABLE 3.8 Change matrix of land use land cover (Sq.km) in 2005 to 2020

Land Class		Land Class 2020								Grand Total	Change %
		BL	**BU**	**CL**	**FL**	**MF**	**SL**	**WB**	**WL**		
Land class 2005	BL	**25.8**	6.84	14.64	7.62	5.26	66.14	0.01	4.68	129.99	56.2
	BU	2.38	**29.86**	13.21	1.09	4.06	1.51	0.3	1.74	53.15	50.2
	CL	61.51	32.25	**7007.56**	39.75	2.05	50.23	7.02	14.68	7216.05	3.3
	FL	29.45	20.18	134.4	**105.94**	1.94	91.03	2.04	19.22	341.2	−25.2
	MF	39.14	1.78	40.38	65.13	**3.45**	31.93	0	2.24	180.05	−89.1
	SL	16.1	7.9	91.88	21.38	1.86	**74.6**	0.11	4.32	218.15	33.6
	WB	129.47	4.36	118.04	14.32	2.52	4.87	**44.5**	9.24	320.32	−83.1
	WL	1.08	5.78	4.8	2.12	1.84	0.53	0.1	**11.41**	27.66	52.5
	Grand total	203.1	106.79	7454.56	255.35	19.68	291.36	54.26	53.02	8486.57	

TABLE 3.9 Change matrix of land use land cover (Sq.km) in 1995 to 2020

Land Class		**Land Class 2020**								**Grand Total**	**Change %**
		BL	**BU**	**CL**	**FL**	**MF**	**SL**	**WB**	**WL**		
Land class 1995	BL	**85.78**	6.84	14.6	7.59	4.29	6.14	0.01	4.68	129.93	56.3
	BU	0.95	**26.84**	9.5	0.35	0.74	0.47	0.29	1.33	40.47	62.1
	CL	60.97	16.18	**6958.85**	21.1	3.17	78.7	7.76	7.45	7152.18	4.2
	FL	23.2	14.18	155.23	**126.38**	6.65	93.52	0.22	6.24	436.62	−41.5
	MF	7.63	1.63	39.42	24.89	**4.45**	87.36	0	2.22	165.24	−88.1
	SL	12.54	17.92	65.11	30.73	1.38	**24.88**	5.11	2.1	227.77	27.9
	WB	13.78	18.58	169.23	24.18	1.45	4.59	**42.56**	21.2	306.57	−82.3
	WL	1.07	5.8	14.78	2.1	0.82	0.54	0.09	**11.3**	27.5	48.1
	Grand total	203.1	106.79	7454.56	255.35	19.68	291.36	54.26	53.02	8486.28	

TABLE 3.10 Change matrix of land use land cover (Sq.km) in 2015 to 2020

Land Class		Land Class 2020									
		BL	**BU**	**CL**	**FL**	**MF**	**SL**	**WB**	**WL**	**Grand Total**	**Change %**
Land class 2015	BL	**125.8**	6.84	444.64	7.62	4.26	36.14	1.01	4.68	660.93	−69.3
	BU	2.38	**79.86**	402.21	1.09	4.06	1.51	1.3	1.74	497.95	−78.6
	CL	40.51	2.25	**4277.56**	33.75	2.05	64.23	3.02	21.68	4603.91	61.9
	FL	9.45	2.18	204.4	**205.9**	1.94	63.03	1.04	17.22	477.35	−46.5
	MF	9.14	1.78	780.38	5.13	**1.45**	129.93	1.03	2.24	844.98	−97.7
	SL	16.1	7.9	931.88	11.38	5.86	**70.6**	0.11	4.32	1210.84	−75.9
	WB	1.47	1.36	1.04	0.32	1.52	4.87	**48.5**	1.24	58.08	−6.6
	WL	1.08	5.78	104.8	2.12	1.84	0.53	1.1	**11.41**	132.53	−59.9
	Grand total	203.1	106.79	7454.56	255.35	19.68	291.36	54.26	53.02	8486.57	

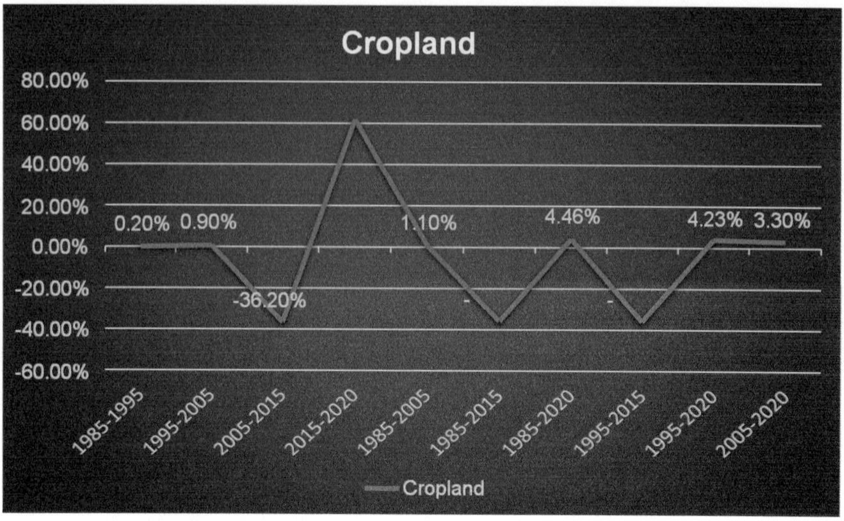

FIGURE 3.3 Change in cropland.

Figure 3.4 shows the change in built-up. It increases from year to year; from 1985 to 1995, there was an increase of 30.7% in the area, from 1995 to 2005, there was an increases of 23.8%, and from 1985 to 2005, there was an increase of 47.3% area. In 2005–2015, there was an increase of 89.3% area, in 1995–2015, an increase of 91.8% area, and in 1985–2015, an increase of 94.4% area. The information in the change matrix shows that the main classes that turn into built-up areas are cropland, barren-land, and waterbody.

Figure 3.5 shows the change in the barren-land. From 1985 to 1995 (2.4%), from 1995 to 2005 (0.1%), and from 1985 to 2005 (2.4%); there was no such significant change in barren-land. In 2005–2015 (80.3%), 1995–2015 (80.3%), and 1985–2015 (80.8%), there is a lot of area increased. Information from the change matrix shows that the main class converted to barren-land is cropland.

Figure 3.6 shows the change in the fallow-land. From 1985 to 1995, the area decreased by 13.9%, from 1995 to 2005 by 21.9%, and from 1985 to 2005 by 32.8%. The area increased by 28.5% in 2005–2015, 8.5% in 1995–2015, and decreased by 5.9% in 1985–2015. From the information in the change matrix, it can be seen that the class in which the converted fallow-land area was maximum is cropland, and the main class that is converted to fallow-land is cropland.

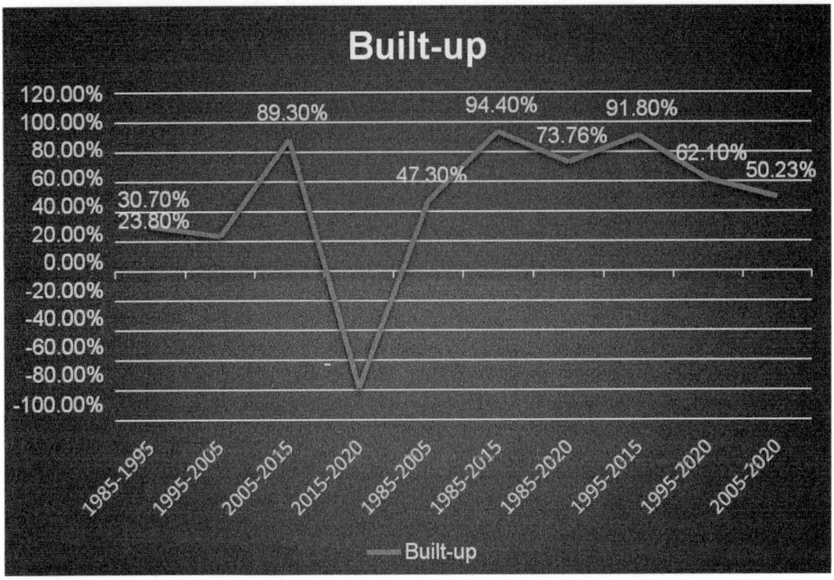

FIGURE 3.4 Change in built-up.

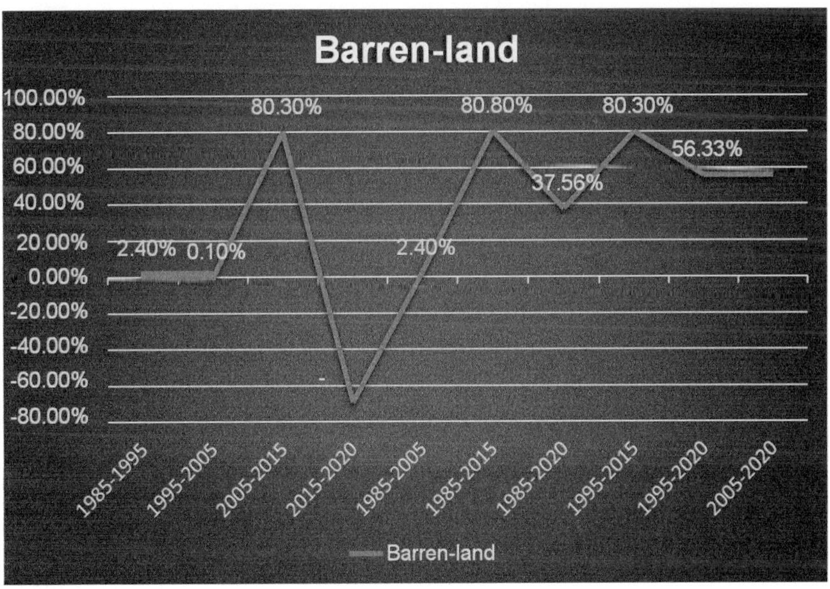

FIGURE 3.5 Change in barren-land.

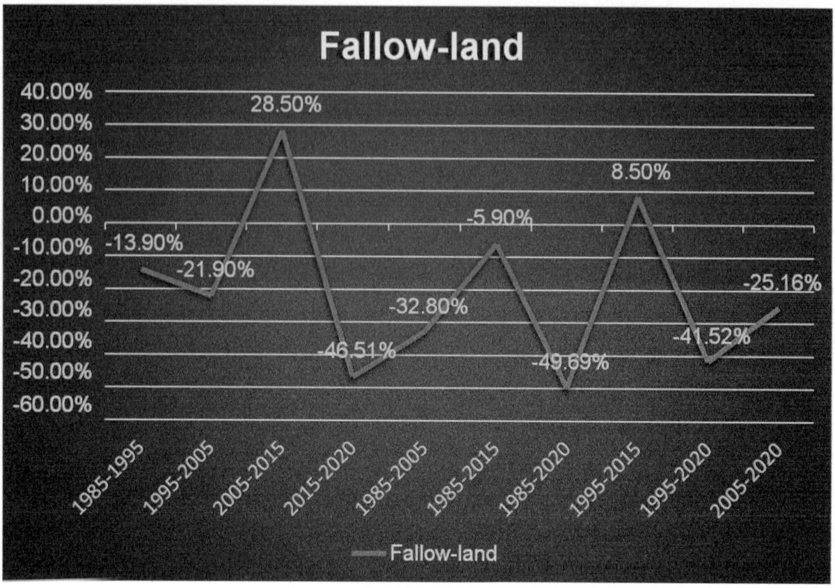

FIGURE 3.6 Change in fallow-land.

Figure 3.7 shows the change of the mixed-forest. From 1985 to 1995, the area decreased by 11.4%, from 1995 to 2005 increased by 8.2%, and from 1985 to 2005 decreased by 3.5%. In 2005–2015, there was an increase in the area of 78.7%, in 1995–2015, there was an increase in the area of 80.4%, and in 1985–2015, there was an increase in the area of 77.9 %. The information in the change matrix shows that the class in which the maximum area of mixed forest conversion was maximum is shrub-land and the main class that is converted to mixed- forest is cropland.

Figure 3.8 shows the change in shrub-land. From 1985 to 1995, the area increased by 9.5%, from 1995 to 2005 (Tables 3.3 and 3.4), it decreased by 4.2%, and from 1985 to 2005, it increased by 5.5%. The area increased by 81.9% in 2005–2015, by 81.2% in 1995–2015, and by 82.9% in 1985–2015. The information in the change matrix shows that the class in which the area converted to shrub-land was maximum is mixed forest and the main class that is converted to shrub-land is cropland.

Figure 3.9 shows the change of the waterbody. From 1985 to 1995, the area increased by 12.4%, from 1995 to 2005 by 4.3%, and from 1985 to 2005 by 16.2%.

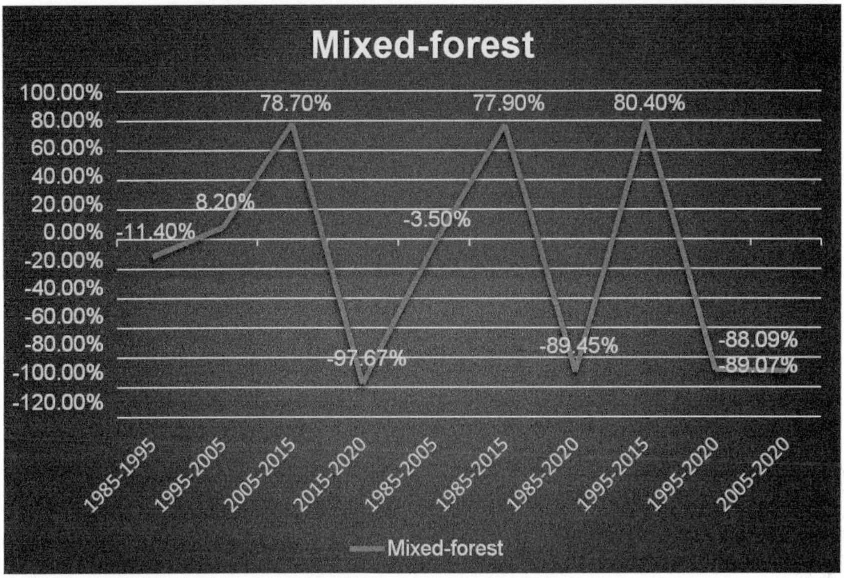

FIGURE 3.7 Change in mixed forest.

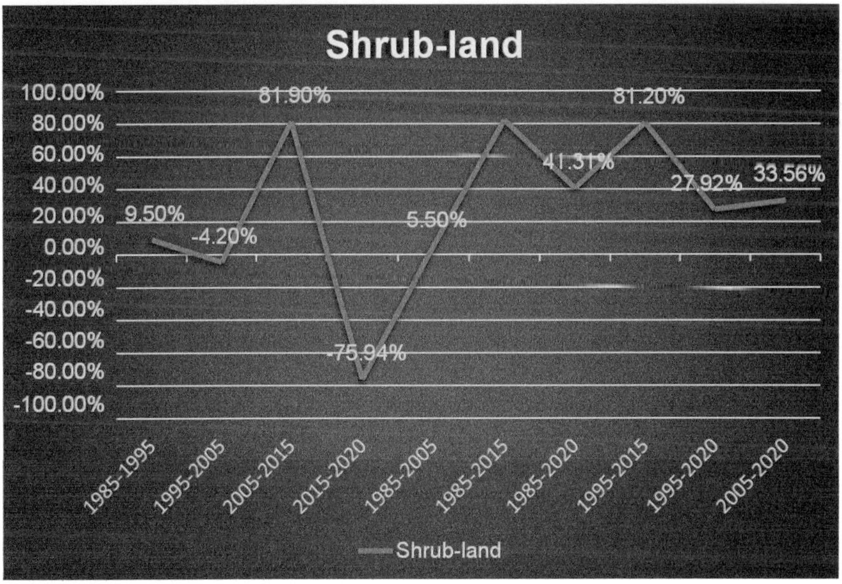

FIGURE 3.8 Change in shrub-land.

In 2005–2015, the area decreased by 81.9%, in 1995–2015 by 81.1%, and in 1985–2015 by 78.4%. The information in the change matrix shows that the class in which the area of waterbody converted was maximum is cropland and cropland is also the main class which is converted into waterbody.

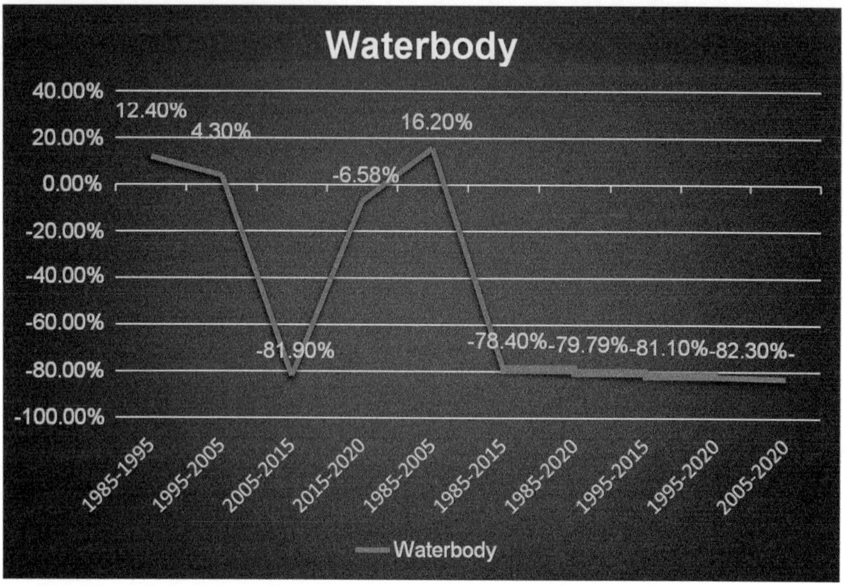

FIGURE 3.9 Change in waterbodies.

Figure 3.10 shows the change in the wetland. From 1985 to 1995, there was an increase in area of 4.2%, from 1995 to 2005 (0.6%), no major change, and from 1985 to 2005, an increase in area of 4.6%. In 2005–2015, the area increased by 79.1%, in 1995–2015, the area increased by 79.2%, and in 1985–2015, the area increased by 80.1%. Information from the change matrix shows that the main class that is converted to wetland is cropland.

3.6 CONCLUSION

Remote sensing and GIS tools have been very useful in analyzing LULC change in Raichur over a period of 30 years. This study will help in the management of the area because the study clearly shows that Raichur is under the threat of rapid urbanization. If we take this trend into account in future, then it is possible that most of the agricultural land would be converted into built-up land. As the built-up area increases, the population will increase and the need of ground water will increase in the future, which would have a direct or indirect impact on cropland. So there is a need to focus on land use and to solve the problems related to the environment. The accuracy of this study is 76.5%. Change analysis shows how each class changes over the time of 30 years. The major classes like cropland, built-up and waterbody show drastic changes. During the period 1985–1995, 1995–2005, and 1985–2005, the cropland areas changed negligibly. However, in 1985–2015, 1995–2015, and 2005–2015 cropland decreased by 35.5%, 35.6%, and 36.2%, respectively. The built-up area continued to increase from 1985 to 2015. During the years 1985–1995, 1995–2005, and 1985–2005, built-up area increased by 30.7%, 23.8%, and 47.3%, respectively. During the time period 1985–2015, 1995–2015, and 2005–2015, the built-up area increased by 94.4%, 91.8%, and 89.3%, respectively.

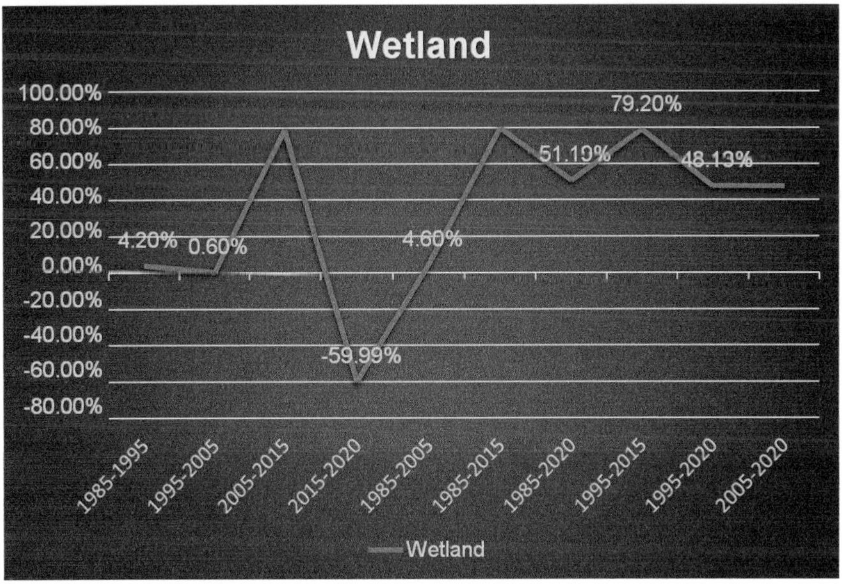

FIGURE 3.10 Change in wetland.

Waterbody show a maximum decrease of 81.9% during the period 2005–2015. The waterbody also decreased by 81.1% and 78.4% in 1995–2015 and 1985–2015, respectively. During the period 1985–1995, 1995–2005, and 1985–2005 waterbody increased by 12.4%, 4.3%, and 16.2%, respectively. The part to be focused is change in the different classes specially in cropland, built-up, and waterbodies during the period 1985–2015. As the population increases, so does the need of water and crops. It can be concluded that the increase in the built- up area leads to a decrease in waterbody and cropland area. This study may be useful in identifying the cropland area which are under risk due to the increase in human settlements.

ACKNOWLEDGMENT

We would like to show my gratitude toward my co-supervisor Dr. Rizwan Ahmad for extending his support all the while. He convincingly kept up my morale and helped me in the completion of my dissertation.

KEYWORDS

- **remote sensing**
- **GIS**
- **LULC**
- **supervised classification**
- **maximum likelihood**

REFERENCES

Zubair and Opeyemi, A. (2006). Change detection in LULC using remote sensing data and GIS (A case study of Ilorin and its environs in Kwara state).

CHAPTER 4

SPATIOTEMPORAL MAPPING OF AIR POLLUTANTS IN DELHI DURING THE HAZE PERIOD

FARHEEN MAHTAB[1], TANZEEL KHATOON[1], and ANAND SHARMA[2]

[1]*Interdisciplinary Department of Remote Sensing & GIS Applications, Aligarh Muslim University, Aligarh, Uttar Pradesh, India*

[2]*Regional Meteorological Centre, Indian Meteorological Department (IMD) Lodi Road, New Delhi, India*

ABSTRACT

Monitoring air quality is of great importance and concern. The study focuses on poor air quality in Delhi during the months of October and November from 2017 to 2019. Five major ground points are taken, namely, Ayanagar, Indira Gandhi International Airport, Delhi Technological University, Lodhi Road, and Netaji Subhas University of Technology Dwarka to cover the whole city. The daily concentration levels of particulate matter, ozone, nitrogen dioxide, and carbon monoxide were used. These levels are averaged and using interpolation shows the overall effects of these pollutants in Delhi around high pollution, which include Diwali and the crop residue burning from the states of Haryana and Punjab.

Sustainable Development: A Geospatial and Statistical Perspective. Qazi Mazhar Ali, Rizwan Ahmad, Irfan Ali, & Prasenjit Chatterjee (Eds.)

DOI: 10.1201/9781779643247-4

4.1 INTRODUCTION

Unpolluted air has so far been treated as an unlimited and free natural resource. However, pure air these days cannot be taken lightly (Chattopadhyay, 2011). When different gases, finely split particles, or finely dispersed liquid aerosols are released into the atmosphere at rates that are greater than the environment's capacity to dissipate, dilute, or absorb them, this is referred to as air pollution. These pollutants may be present in the air at levels that have detrimental impacts on one's health, finances, or aesthetics.

Water vapor, which typically ranges from 0.01% to 4% by volume and can reach as high as 5% in extremely humid conditions, is another typical, but highly variable, component of the atmosphere. Sulfur dioxide, nitrogen dioxide, and carbon monoxide are the gaseous air pollutants that are most dangerous in cities because they are released into the atmosphere when fossil fuels like fuel oil, gasoline, and natural gas are burned in power plants, vehicles, and other combustion sources. In the atmosphere, ozone, another gaseous pollutant, is created through complicated chemical interactions between nitrogen dioxide and a variety of volatile organic molecules, such as gasoline vapors. Due to their severely detrimental effects on human health, "particulates" in the air, such as soot, dust, smoke, fumes, and mists, are classified as major air pollutants when they have a diameter of fewer than 10 μm (m; millionths of a meter). Different industrial operations, coal- or oil-burning power plants, home heating systems, and autos all release them (https://www.britannica.com/print/article/10772). A significant contaminant of many diesel fuels is lead vapors, which are highly hazardous and less than 0.5 m in size. The majority of these compounds are mostly safe when found in low quantities in nature; they only turn into pollutants when their concentrations are significantly higher than the background value and start to have negative effects. Pollutants are mostly introduced by human activity, especially industrial manufacturing and driving. However, some air pollutants may be introduced by natural events like volcanic eruptions, wind-driven soil erosion, forest fires, sand storms, and plant pollen dispersal (UNCHS, 1996). Nearly half of the world's population is predicted to live in cities and other metropolitan regions, where these activities are mostly concentrated. Delhi is the capital of India and has already been declared the pollution capital of the world. So, it is very necessary to keep a check on air quality in order to delineate appropriate air quality management plans; quantification of emissions from different air pollution sources and their impact on ambient air quality become essential. So, before stepping into any management strategy for reducing air pollution synoptic and systematic monitoring of ambient air quality is a need.

4.2 STUDY AREA

Delhi, India's capital city, is located in the country's center north. It, the National Capital Territory, occupies an area of 1483 square kilometers and is located in Northern India at 28.61°N 77.23°E. Delhi, one of the largest cities in the country, is located largely on the west bank of the Yamuna River, a tributary of the Ganges River. It is situated between 213 and 305 m above the sea level. The National Capital Territory includes the adjacent metropolitan region as well as the surrounding rural areas. Due to its historical significance as India's political, commercial, transportation, and cultural capital, as well as its high population density, Delhi is particularly vulnerable to declining air quality (Figure 4.1).

4.3 THE GLOBAL REACH OF AIR POLLUTION

Air pollution is a global problem that travels enormous distances due to winds and lingers in the atmosphere, having no local, regional, or continental boundaries. Temperature and weather patterns on Earth are also influenced by air pollution. Due to the increasing use of fossil fuels worldwide, the amount of carbon dioxide in the atmosphere has been steadily increasing since 1900, and this rise is occurring at a faster rate. By the end of the twenty-first century, an extra 4°C (7.2°F) of worldwide air temperature increase is predicted if carbon dioxide levels are not lowered. This rising trend may lead to the melting of the polar ice caps, an increase in sea level, and flooding of coastal areas globally. In certain parts of the world, increased heat and humidity might increase the likelihood of disease transmission to humans and animals due to modified precipitation patterns caused by global warming. International accords aimed at lowering greenhouse gas emissions must be put into practice in order to preserve the quality of the world's air and lessen the consequences of global warming (https://www.britannica.com/print/article/10772). The main source of air pollution is urban areas. Globally, urban air pollution

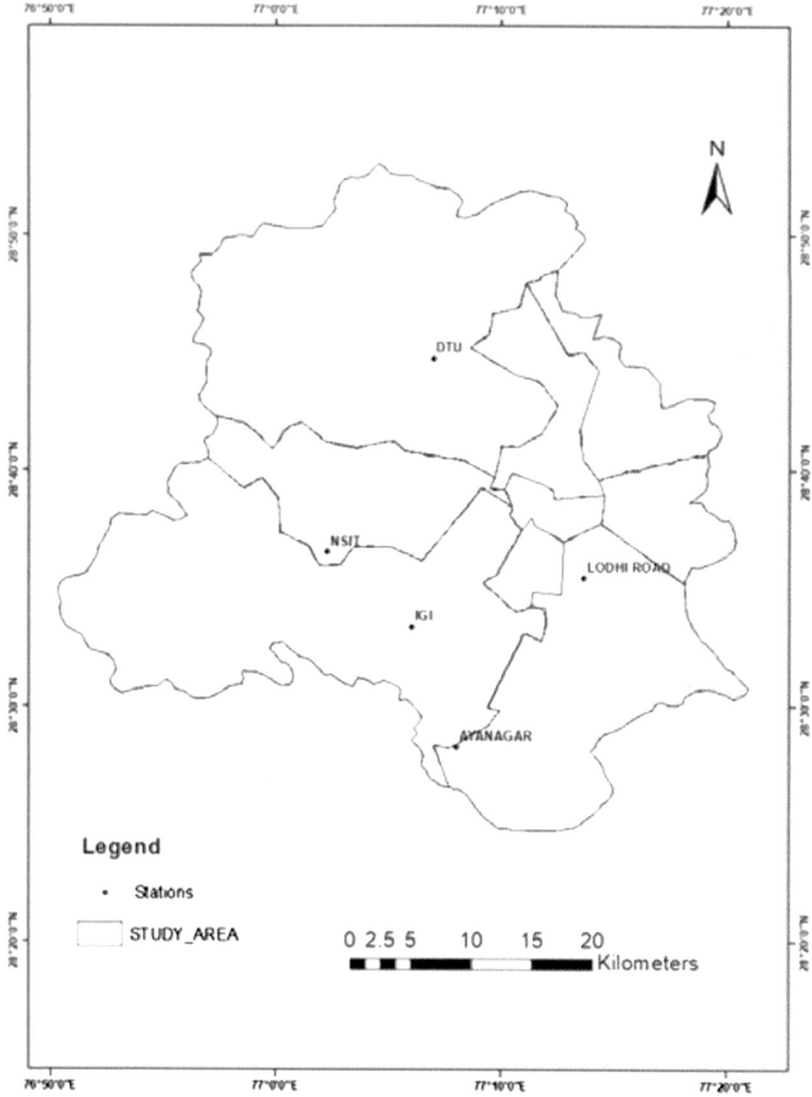

FIGURE 4.1 Study area.

poses a serious threat to the environment and public health. The World Health Organization (WHO) estimated that ambient air pollution was responsible for 3.7 million deaths globally in 2012, or around 6.7% of all fatalities (Burden of Disease from Ambient Air Pollution for 2012).

The extent of the air pollution issue is growing as cities become more populated. As a result, living in an urban environment is now accompanied by pollution. The quality of air in cities has significantly declined as a result of urbanization, industrialization, and economic expansion (Wahid, 2006). Particulate and gaseous pollutants were released into the atmosphere as a result of the increasing usage of fossil fuels and their derivatives brought about by modernization and increased industrial activity, especially in emerging nations. The transportation industry is the primary source of pollution in metropolitan areas, generating NO_2, Pb, and CO. Pollution is also increased by industry, power plants, and the burning of solid waste. In many parts of the world, uncontrolled urbanization and rapid economic expansion are leading to severe air pollution-related issues. According to estimates from the WHO, 1.4 billion people who live in cities in developing nations breathe air that is more polluted than what is considered safe for their health (UNEP/WHO, 1992). Human health is significantly impacted by both indoor and outdoor sources of air pollution, particularly in poor nations. Indoor causes of air pollution include tobacco smoke and the burning of solid fuels used for cooking and heating (Bruce et al., 2000). However, the study's main focus is on ambient air pollution in cities. Owing to the high levels of ambient air pollution and its possible impacts on a population of around 16.7 million people, this research is being conducted in New Delhi, India. Nonetheless, the findings are useful for controlling the quality of the air in other Indian towns and other places.

4.4 AIR POLLUTION IN INDIA

In recent times, uncontrolled urbanization and fast industrial expansion have led to a steady decline in the quality of air in Indian cities (MEFCC, 2009). Numerous scholars have investigated the connection between environmental deterioration and development at the national level. India's megacities have seen a sharp increase in both urban population and population density due to migration from rural to urban regions. Due to their closeness to industrial and transportation sources, densely populated areas are also more likely to have higher air pollution levels, which expose a larger population to higher air pollution levels. In metropolitan India, the main causes of air pollution have been found to be motor vehicles, power plants, industry, and biomass burning (Kandlikar and Ramachandran,

2000). The primary causes of air pollution in rural and urban settings differ significantly, as would be predicted. Using solid fuels for cooking is a major cause of air pollution exposure in rural India. Solid fuels are used for cooking in over 86% of Indian rural homes compared to just 26% of urban households. Open-field burning of agricultural residue is another significant source of particulate matter (PM) in rural areas, and it has a proven regional influence on air quality (Badarinath et al., 2009). Merely 3% of homes in New Delhi cook with solid fuels (Census of India, 2011). While ambient air pollution in urban surroundings, particularly New Delhi, is a major health problem in rural India due to the extensive use of solid fuels for cooking, this research only looks at ambient air pollution indoors.

4.4.1 AIR POLLUTION IN NEW DELHI, INDIA

India's capital, New Delhi, is a megacity with 18.4 million people living there as of the end of 2019. In 2011, there were 16.7 million people living there. Population density thus sharply dropped. The Indian economy has been booming at the same time as this population expansion, particularly in its big cities. As earnings have increased, New Delhi's air quality has been worse. In New Delhi, there are 2.8 million automobiles, 5.7 million two-wheelers, 81,000 three-wheelers, 79,000 taxicabs, and 20,000 buses. The motorized vehicle industry is growing at a pace of around 7% annually. The transportation sector contributed the most (17%) to regional PM2.5 emissions for New Delhi and its surrounding cities, followed by power plants (16%), brick kilns (15%), industries (14%), household sources (12%), garbage burning (8%), and others (Pant et al., 2018).

New Delhi now has some of the highest air pollution levels among the world's major cities.

4.5 DATA SETS AND METHODOLOGY

Pollution concentration is evaluated using data from the Central Pollution Control Board (CPCB) website. The Ministry of Environment and Forests of the Central Government oversees the CPCB, which is tasked with gathering information on air pollution and setting guidelines for air quality.

4.5.1 DATA PROCESSING

To investigate the amount of pollution at various sites, an improved tabular table obtained from CPCB is utilized in conjunction with the interpolation approach. Inverse distance weighted (IDW) is one of the interpolation techniques utilized.

4.5.2 INTERPOLATION METHODS

The process of guessing values at additional unknown sites by using sample points or points with known values is known as interpolation. Unknown values can be forecasted using any geographic point data, including elevation, rainfall, chemical concentrations, noise levels, etc.

4.5.2.1 INVERSE DISTANCE WEIGHTED

The foundation of the inverse distance weighting interpolator is the notion that each input point has a local effect that decreases with distance. It gives points nearer the processing cell more weight than ones farther away. The output value of each site may be calculated from a predetermined set of points or from all the points within a certain radius. IDW refers to the fact that points closer to the forecast location are given more weight than points farther away. Each grid node's value is determined using the IDW method by looking at nearby data points that are within a user-specified search radius. The interpolation procedure can make use of any or all of the data points. The weighted total of all the points is averaged to determine the node value. Data points that are increasingly farther away from the node have far less impact on the calculated value than do data points that are closer to the node (Figure 4.2).

After choosing the data points to utilize for the computation, a radius is created around each grid node. Power, fixed, variable, and search radius controls, as well as a barrier, are available to regulate how IDW is used (Figure 4.3).

Figure 4.4 represents the daily concentration of PM2.5 for the months of October and November 2019. The concentration of PM was found to be considerably high, with their mean ranging from 141.36 to 183.18. A spike in the concentration of pollutant can also been seen after Diwali which is October 27 (Figure 4.5).

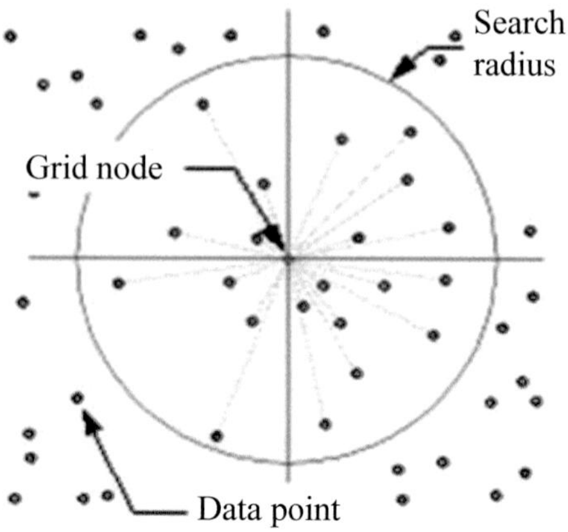

FIGURE 4.2 Interpolation.

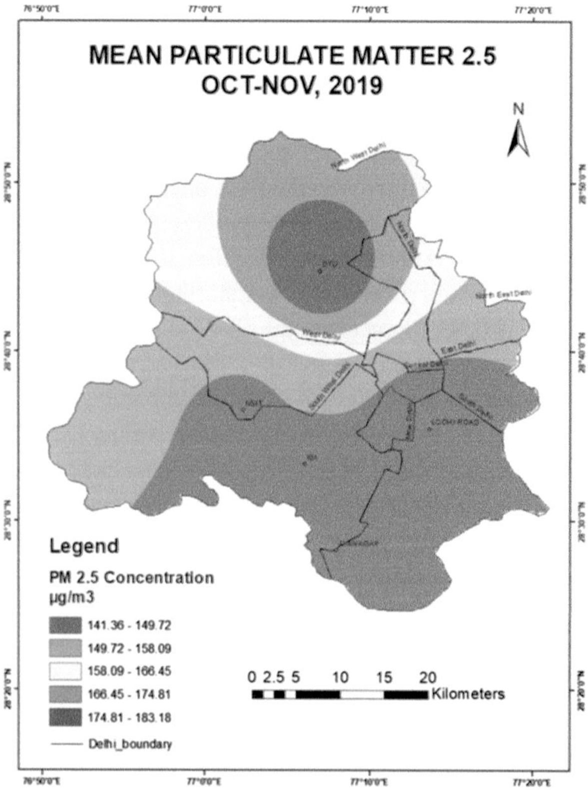

FIGURE 4.3 Map showing PM2.5 mean in Delhi.

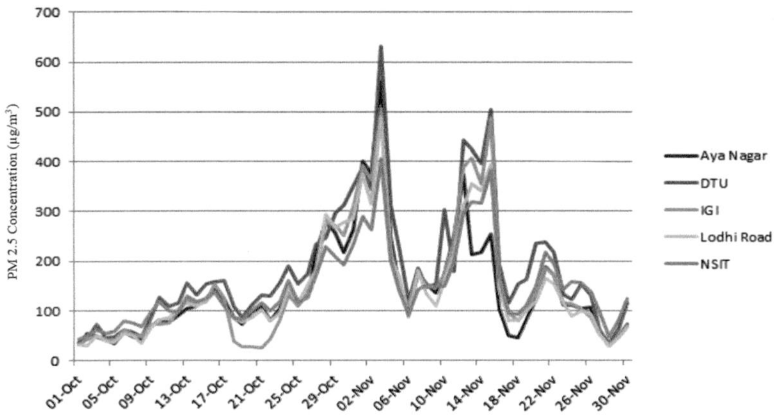

FIGURE 4.4 Linear representation of daily PM2.5 concentration in Oct–Nov 2019.

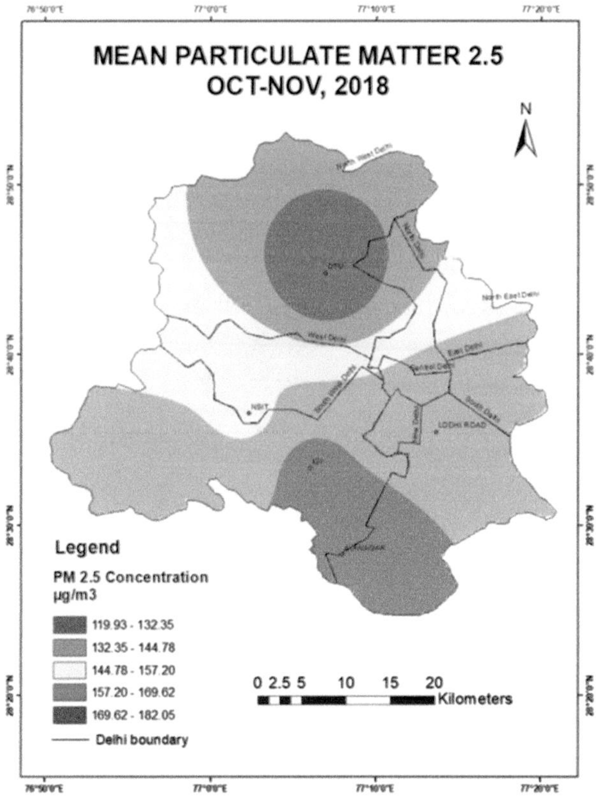

FIGURE 4.5 Map showing PM2.5 mean in Delhi.

Figure 4.6 represents the daily concentration of PM2.5 for the months of October and November 2018. The concentration of PM was found to be considerably high in 2018 but slightly lower than that of 2019, with their mean ranging from 119.93 to 182.05. A spike in the concentration of pollutants can also be seen after Diwali which is November 7 (Figure 4.7).

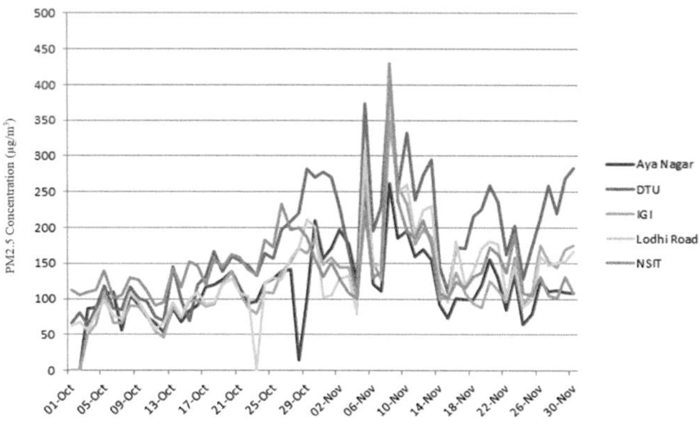

FIGURE 4.6 Linear representation of daily PM2.5 concentration in Oct–Nov 2018.

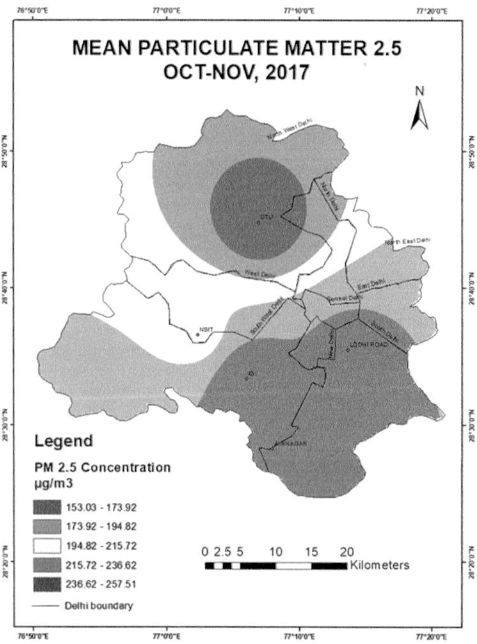

FIGURE 4.7 Map showing PM2.5 mean in Delhi.

Figure 4.8 represents the daily concentration of PM2.5 for the months of October and November 2017. The concentration of PM was found to be considerably high for 2017 higher than that of 2018 and 2019, with their mean ranging from 153.03 to 257.51. A spike in the concentration of pollutants can also be seen after Diwali which is October 19 (Figure 4.9).

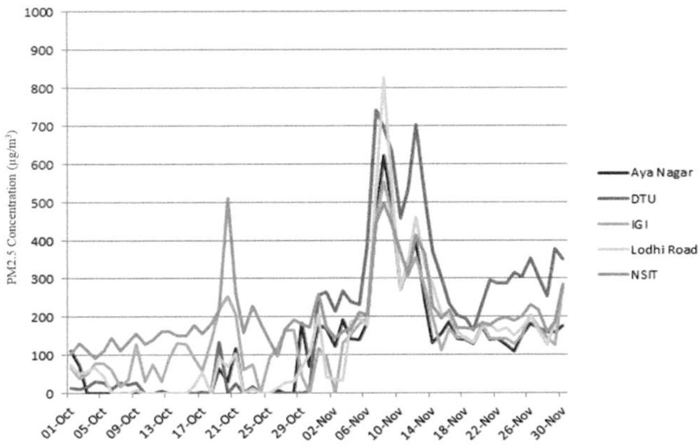

FIGURE 4.8 Linear representation of daily PM2.5 concentration in Oct–Nov 2017.

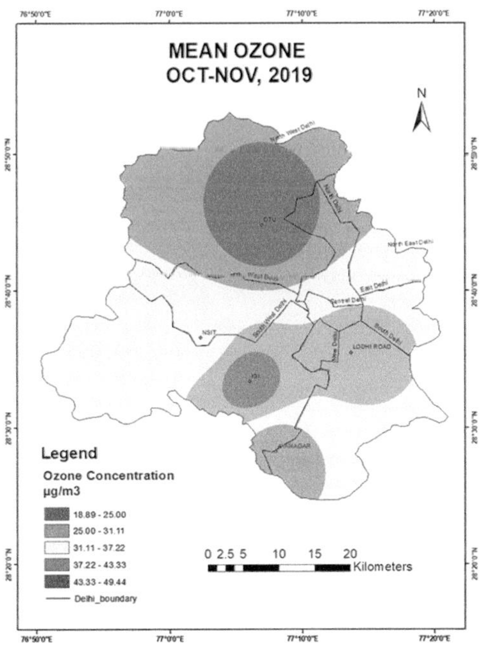

FIGURE 4.9 Map showing ozone mean in Delhi.

Figure 4.10 represents the daily concentration of ozone for the months of October and November 2019. The concentration was found to be considerably high for 2019, with their mean ranging from 18.89 to 49.44. A spike in the concentration of pollutants can also be seen after Diwali which is October 27 (Figure 4.11).

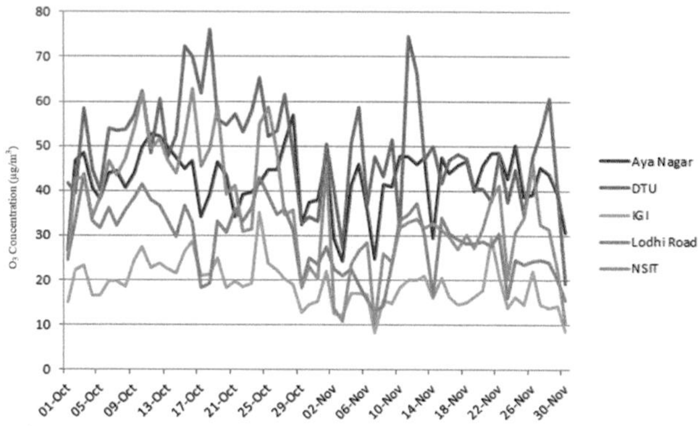

FIGURE 4.10 Linear representation of daily ozone concentration in Oct–Nov 2019.

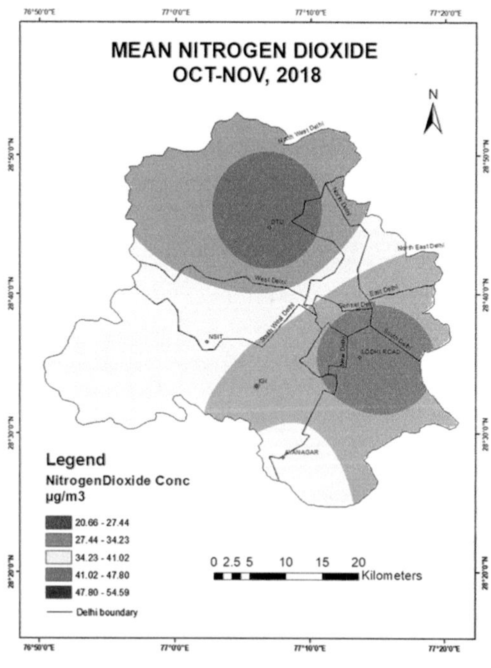

FIGURE 4.11 Map showing NO_2 mean in Delhi.

Figure 4.12 represents the daily concentration of Nitrogen dioxide for the months of October and November 2018. The concentration was found to be considerable, with their mean ranging from 20.66 to 54.59. A spike in the concentration of pollutant can also be seen after Diwali which is November 7 (Figure 4.13).

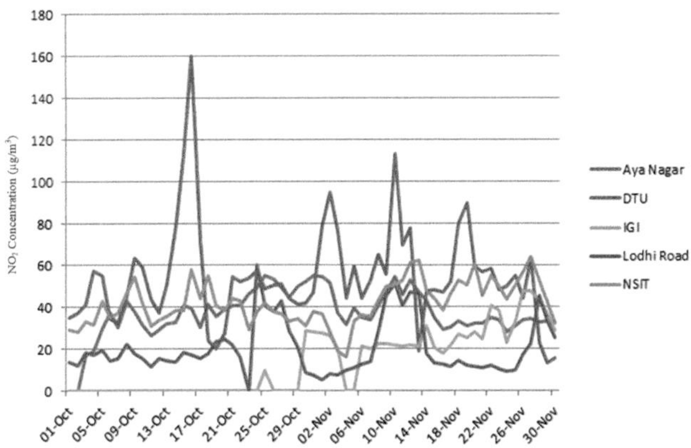

FIGURE 4.12 Linear representation of daily NO_2 concentration in Oct–Nov 2018.

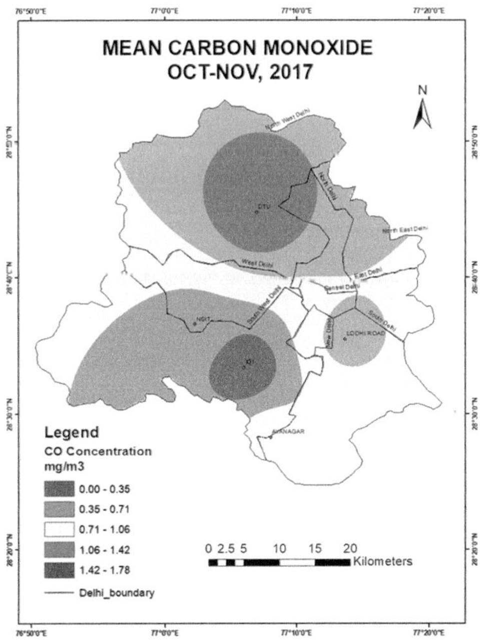

FIGURE 4.13 Map showing CO mean in Delhi.

Figure 4.14 represents the daily concentration of carbon monoxide for the months of October and November 2017. The concentration was found to be considerable, with their mean ranging between 0 and 1.78 in mg/m^3. Spikes in concentration at the end of November are indicative of crop residue burning (CRB) from the surrounding states mainly Punjab and Haryana.

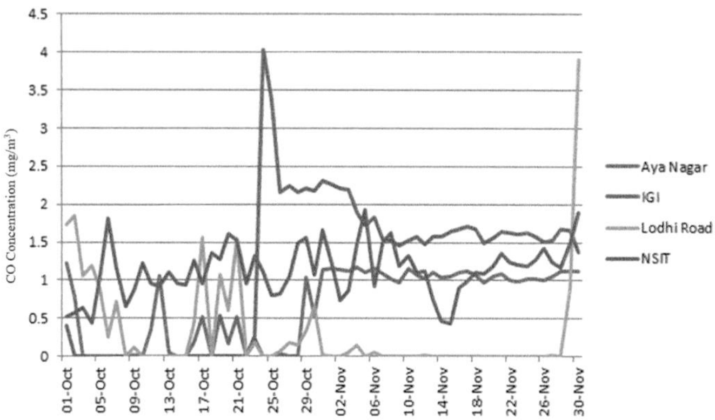

FIGURE 4.14 Linear representation of daily CO concentration in Oct–Nov 2017.

4.6 CONCLUSION

A summary of air quality in the research region is given in this publication, along with recommendations for preserving it. Ayanagar, Delhi Technological University, Indira Gandhi International Airport, Netaji Subhas University of Technology, and Lodhi Road make up the majority of the research area, which is dedicated to examining Delhi's general air quality. To determine the extent of the city's pollution level, these five primary places are investigated. The impact of climatic conditions on the dispersion of air contaminants is a well-established phenomenon. The influence of postmonsoon CRB may be the cause of the increased concentration of pollutants in October and November. Most likely, CRB, which is common in Punjab and Haryana throughout the winter, contributes to the pollution caused by burning biomass. Because of the ground-based inversion, low wind conditions, and cold, dry air that dominate the winter months in Delhi, the concentration of pollutants increases at this time of year. A layer of warm air acts as a cap on top of this layer, trapping high concentrations of pollutants near the earth's surface Furthermore, heavy smog production throughout the winter has also been observed in Delhi; this is caused by both vehicle pollution and the weather that prevails in December and January. An analysis indicates that the Indo-Gangetic plains burn about 500 million tons of agricultural waste in the fall and winter. Delhi has considerable smog development throughout the winter months due to the wind blowing from the north and north-west of India toward the east. This wind finally combines with pollutants. The months of October–November are particularly affected. Although the exact causes of the recent drop in air quality are unknown, a warmer climate, an increase in wildfires, and changing patterns of human consumption brought on by population expansion and a robust economy may all be to blame. Furthermore, the long-term picture is yet unknown. In the present study multiple air pollutants modeling using the Interpolation technique along with remote sensing and geographic information system (GIS) is used to map the pollution capital. In this study, the five most common pollutants, namely, PM10, PM2.5, ozone, nitrogen dioxide, and carbon monoxide are considered. Each pollutant's daily concentration was recorded and maintained by CPCB is shown through the graph. The daily concentrations of these pollutants are averaged, and after combining these parameters, IDW map was created to model the quality and extent of these pollutants. The study shows that the pollution levels were high during the months of October and November due to several factors such as metrological conditions, CRB, and also due to inappropriate burning of crackers during Diwali celebrations. The concentration of the pollutants was found to be high during 2019 as compared to 2017. In some cases, due to improper monitoring, concentration levels were very poor in 2017. Proper rules and awareness have led to control the pollution to a great extent. The government of the capital has taken some major steps to combat levels of pollution in the city that may not only combat the deficit of poor air but also contribute largely to help to reduce the climatic changes globally. In reducing pollution, the government and the working bodies encourage the community or individuals to avail and practice more clean and sustainable options on a daily basis. The work highlights the potential for future analysis using GIS and remote sensing techniques to automate inventories and improve the air quality of the urban areas. However, the resulting interpolation map of the study area can be used by various authorities, researchers, scientists, or government for maintaining proper measurements, techniques, and plans for various development activities/schemes in those regions in order to build a better, safer, and sustainable environment for all.

ACKNOWLEDGMENT

We would like to thank Dr. QaziMazhar Ali, Chairman of IDRS and GIS Applications, A.M.U. for encouraging me. We would like to express my special thanks of gratitude to the Director IMD and mentor, Dr. Anand Sharma who gave me the golden opportunity to do my work at IMD, and we would also like to thank Ms Palak Sharma Scientific Assistant for her support. We also show my special gratitude to Dr. Rizwan Ahmad for being inspiring, supportive, and humble during my entire work. We are also thankful to my friends and family for their moral support.

KEYWORDS

- **crop residue burning**
- **interpolation**
- **metrological**
- **pollutants**
- **spatiotemporal**
- **thematic**

REFERENCES

Air Pollution—Britannica Online Encyclopedia. Retrieved from https://www.britannica.com/print/article/10772

Badarinath, K. V. S., Kumar, K. S., and Sharma, R. (2009). A. Long-range transport of aerosols from agriculture crop residue burning in Indo-Gangetic Plains—A study using LIDAR, ground measurements and satellite data. *Journal of Atmospheric and Solar-Terrestrial Physics 71*, 112–120.

Bruce, N., Perez-Padilla, R., and Albalak, R. (2000). Indoor air pollution in developing countries: A major environmental and public health challenge. *Bulletin of the World Health Organization 78*, 1078–1092.

Burden of Disease from Ambient Air Pollution for 2012. Retrieved from http://www.who.int/phe/health_topics/outdoorair/databases/AAP_BoD_results_March2014.pdf?ua=1 (Accessed March 1, 2015).

Chattopadhyay, S. (2011). *Spatial and Temporal Variations of Ambient Air Quality in Burdwan Town*, West Bengal, India.

Census of India; 2011. Retrieved from http://www.censusindia.gov.in/2011census/population_enumeration.html (Accessed July 16, 2015).

Kandlikar, M., and Ramachandran, G. (2000). The causes and consequences of particulate air pollution in urban India: A synthesis of the science. *Annual Review of Energy and the Environment 25*, 629–684.

Molina, M. J., and Molina, L. T. (2004). Mega cities and atmospheric pollution. *Journal of the Air and Waste Management Association 54*, 644–680.

Rejeski, D. (1993). GIS and risk: A three-culture problem. In: Goodchild, M. F., Parks, B. O., Steyaert, L. T. (Eds.). *Environmental Modelling with GIS*. NY: Oxford University Press, New York, pp. 318–331.

Saraswat, A. (2015). Air pollution in New Delhi, India: Spatial and temporal patterns of ambient concentrations and human exposure.

Sarath, K., Guttikunda, R. G., and Pant, P. (2014). Nature of Air Pollution, Emission Sources, and Management in the Indian Cities.

Sharma, M. (2016). Comprehensive Study on Air Pollution and Green House Gases (GHGs) in Delhi.

UN. (1992). In: *Agenda 21: Rio Declaration, Rio de Janeiro: Final Text of the Agreement Negotiated by Governments of the United Nations Conference on Environment and Development*.

UNCHS. (1996). *An Urbanizing World: Global Report on Human Settlements, 1996*. Oxford University Press.

Wahid, A. (2006). Productivity losses in barley attributable to ambient atmospheric pollutants in Pakistan. *Atmospheric Environment 40*, 5342–5354.

CHAPTER 5

MAPPING THE GROUNDWATER TABLE IN AN ALLUVIAL AQUIFER USING DIGITAL ELEVATION MODEL, GIS, AND GEOSTATISTICS

HARIS HASAN KHAN[1], ARINA KHAN[2], GHAZALA SIRAJ[3], and TAQVEEM ALI KHAN[3]

[1]*Interdisciplinary Department of Remote Sensing and GIS Applications, Aligarh Muslim University, Aligarh, Uttar Pradesh, India*

[2]*Department of Geological Sciences,George Washington University, Washington, D.C., USA*

[3]*Department of Geology, Aligarh Muslim University, Aligarh, Uttar Pradesh, India*

ABSTRACT

Groundwater table elevation maps are essential for understanding hydrogeological processes, including groundwater flow, recharge, and interactions with surface water. Traditional interpolation techniques often fail to accurately reproduce the water table's spatial variability due to sparse well data and the neglect of topographic influences. This study evaluates various interpolation methods—Inverse Distance Weighted (IDW), Thin Plate Spline, Ordinary Kriging, and Kriging with External Drift (KED) to map groundwater elevations in the alluvial aquifers of western Uttar Pradesh, India, using water table elevation data from 150 observation wells.

Sustainable Development: A Geospatial and Statistical Perspective. Qazi Mazhar Ali, Rizwan Ahmad, Irfan Ali, & Prasenjit Chatterjee (Eds.)

DOI: 10.1201/9781779643247-5

While IDW and Thin Plate Spline interpolations exhibited anomalies, such as unrealistic water table heights, Ordinary Kriging marginally improved results but lacked consistency with topographic trends. The KED method, incorporating DEM as an external drift predictor, demonstrated the most reliable performance by closely aligning the groundwater table with surface topography. However, the model did not fully capture expected interfluve and floodplain behaviors, suggesting the need to integrate additional parameters. This study underscores the value of KED in enhancing groundwater mapping accuracy and provides insights for groundwater management agencies to adopt advanced interpolation techniques.

5.1 INTRODUCTION

Water table elevation maps are fundamental elements of hydrogeological investigations at regional scales. These maps are used to assess the groundwater flow directions, groundwater recharge or discharge zones, groundwater–surface-water interactions, and the effect of anthropogenic and natural stresses on the groundwater system. Mapping of the groundwater table is usually conducted by obtaining water level data in groundwater wells and then by interpolating the water levels to create a continuously varying surface representing the groundwater table. However, the algorithms used for interpolating water level data do not reproduce the actual shape of the water table. This is usually due to the sparse nature of the well data as well, as they do not consider the relationship of the water table with the surface topography.

A close relationship exists between the groundwater table and surface topography. In other words, the groundwater table is a subdued replica of the surface topography. This implies that topographic data can assist in mapping water table elevations, which is otherwise based on sparse water level data. Mainly, Digital Elevation Models (DEMs) offer a comprehensive spatial characterization of land surface geometry and stream networks that can be used as a secondary variable to improve the estimation of the sparsely sampled primary variable, that is, the water table. This is done by a technique known as "kriging with an external drift" or Kriging with External Drift (KED) (Marechal, 1984; Ahmad and de Marsily, 1987; Chiles, 1992; Deutsch and Journel, 1992; Goovaerts, 1997). KED maps the primary variable (groundwater table) at all nodes of the estimation grid by adding the stationary stochastic residual and the deterministic trend function obtained from the secondary variable (DEM). Originally external drift was implied as a smoothly varying spatial function. Later, it was amended to incorporate any deterministic function explicitly specified by its grid node values and linearly linked (post-transformation, if required) to the primary regionalized variable. DEM elevations provide an ideal external drift function for mapping spatially variable entities controlled by surface topography. Raster-DEMs are available at a uniform spatial resolution and discretized on a regular grid. DEM-derived external drift functions have been used with Kriging for temperature mapping (Hudson and Wackernagel, 1994), rainfall mapping (Goovaerts, 2000), soil moisture mapping (BaÂrdossy and Lehman, 1998), and soil thickness mapping (Bourennane et al., 1996).

At regional scales, groundwater table elevations generally show nonstationary behavior. This is not in agreement with standard assumptions of cokriging. There are valid physical reasons to hypothesize that linear connections exist between water levels and DEM variables that meet the assumptions of the external drift model. This investigation examines a connection of this nature, where the elevation of the water table is represented as a linear function of the topographic elevation. The KED model is then illustrated by mapping water table elevations in the alluvial aquifer in Aligarh and neighboring districts of Uttar Pradesh in India.

This work aims to evaluate the effectiveness of various interpolation techniques for interpolating groundwater elevation data. The aim is to emphasize the best technique groundwater agencies should employ to prepare water table maps.

5.2 STUDY AREA

The study area (Figure 5.1) is a rectangular region covering New Delhi, several districts (Aligarh, Agra, Mathura, Meerut, Bulandshahr, Moradabad, Firozabad, Mainpuri, Etah, Ghaziabad, Budaun, Baghpat, Gautam Budha Nagar, Jyotiba Phule Nagar of western Uttar Pradesh, and Gurgaon and Bayana districts of Haryana and Rajasthan States, respectively.

The region is characterized by a monsoon climate with a markedly wet season from June to September, during which most of the rain is received and the aquifers are recharged. Temperature ranges from close to 1°C during December–January to a maximum of around 44°C during May–June in many parts of the region.

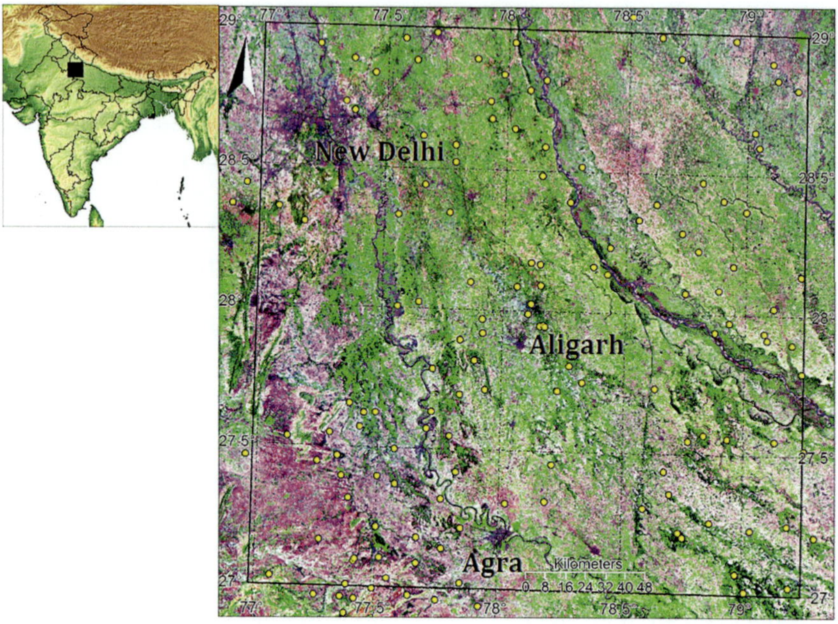

FIGURE 5.1 Location of the study area. (*Yellow* dots indicate the location of CGWB observation wells used in the present study.)

The region is a depositional basin for Himalayan sediments and is characterized by very low relief. The area is drained by rivers Yamuna and Ganga and their tributaries. The general drainage direction is northwest to southeast. Geologically, the area consists of older alluvium (Bhanger) mainly found in the interfluvial regions, and younger alluvium (Khadar) mainly occurring in the floodplains. The area is underlain mainly by alluvial aquifers (Figure 5.2) of unconfined to semiconfined conditions. A three-tier aquifer system is present in most of the locations. The topmost aquifer in the region is, however, in fully unconfined conditions.

5.3 DATA AND METHODOLOGY

For the present work, groundwater depth data from 150 CGWB groundwater wells (for the premonsoon season of 2016) within the study area was downloaded from the India Water Resource Information System website (http://www.india-wris.nrsc.gov.in). A vector (shapefile) file of well locations with water table depth attributes was prepared in SAGA GIS software. Surface elevation at well locations was obtained from 30-m Shuttle Radar Topography Mission (SRTM) digital elevation model (DEM) available from the website, https://earthexplorer.usgs.gov/, using an automated attribute extraction process in SAGA GIS.

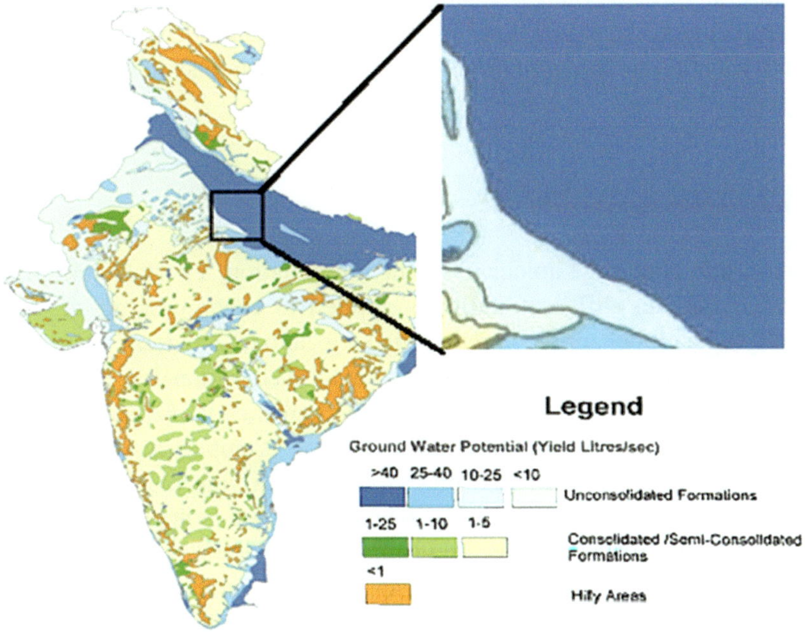

FIGURE 5.2 Location of study area on the hydrogeological map of India.

Source: Central Ground Water Board [CGWB]

Groundwater table elevations were obtained by subtracting the depth from water table values obtained from CGWB data from the surface elevation values obtained from SRTM DEM. The water table elevation data at the 150 well locations were then interpolated using several interpolation methods. The present study considers the following interpolation methods (a brief account of each is given below):

1. Inverse Distance Weighted (IDW)
2. Thin Plate Spline
3. Ordinary Kriging
4. KED

5.3.1 INVERSE DISTANCE WEIGHTED (IDW)

In IDW interpolation, the estimation of the interpolated value is done using the weighted mean of the data, and the weights are inversely proportional to the squared distance between the estimated value and each data value. This leads to a smooth surface between the data points, but a discontinuity in slope at the data points is created. For unevenly spaced hydrologic data, this method produces artifacts. Hydrological variables do not usually vary in the way this method assumes. Since the method is easy to use, it is quite popular.

5.3.2 THIN PLATE SPLINE

Within thin plate (or Laplacian) spline interpolation, a surface that maintains continuous differentiability is applied to encompass all the data, signifying a global approach rather than a local one. The term "thin plate" originates from the minimizing function employed, which draws a physical comparison to the average curvature or flexural energy of a slender elastic sheet. Typically, a polynomial of modest degree is employed, and adhering to a minimum curvature criterion involves identifying the function with the utmost smoothness attainable. Consequently, this approach resolves the challenges of overfitting or oscillation encountered in the moving polynomial method. Two primary adaptations of this technique exist. The more straightforward version presupposes error-free data, causing the interpolated surface to intersect the data points. It incorporates a singular "smoothing" or "tension" parameter intended to govern the smoothness of the interpolated surface. The alternate variation accommodates measurement errors by including an extra parameter that signifies the error term. This parameter regulates the equilibrium between surface smoothness and the proximity of the interpolated surface to the data. Thin plate splines demonstrate exceptional effectiveness across most hydrologic applications unless the data is substantially more dispersed than the scale of the underlying phenomenon. They exhibit resilience and practical simplicity in their application. This technique is typically endorsed within the deterministic interpolation framework. When contrasted with geostatistical techniques like Ordinary Kriging (OK) and KED, the sole drawback is the relatively less straightforward incorporation of explicit consideration for measurement and estimation errors within spline interpolation.

5.3.3 *ORDINARY KRIGING*

OK (as well as KED) is based on the notion that the interpolated pattern is a random variable, which can be described by the variogram. The variogram is the variance between pairs of points spaced at a certain distance (or lag), or equivalently a measure of spatial correlation. Pairs of points that are close to each other tend to be similar; hence, the value of the variogram (termed gamma) at small lags tends to be small. As the lag increases, so does gamma, as the values become increasingly dissimilar at larger lags. Figure 5.3a shows a typical variogram of a stationary pattern (i.e., a pattern for which the mean does not change with space).

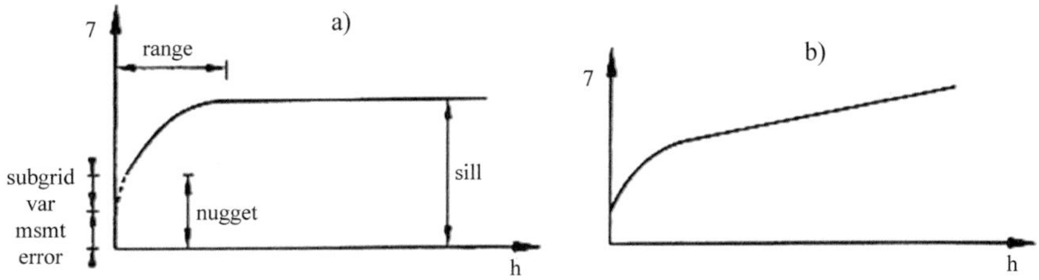

FIGURE 5.3 (a) Variogram of a stationary random function (i.e., no spatial trend) showing a nugget consisting of subgrid (small scale) variability and measurement error, sill, and range; (b) variogram for a random function that is nonstationary in the mean (i.e., possesses a spatial trend).

Three parameters need to be specified for a variogram, two of which are the sill (which is the spatial variance of the overall pattern) and the range (which is the spatial correlation length and a measure of how continuous or smoothly varying the patterns are—large ranges relating to smoothly varying patterns, short ranges relating to erratically varying patterns). If the pattern is not stationary in the mean (i.e., exhibits a spatial trend), it neither possesses a sill nor a range (Figure 5.3b), but it may still be used for geostatistical analyses. The third parameter is the nugget (Figure 5.3a), which represents the variance of pairs of points at the same location. Each of these parameters, the sill, range, and nugget can be interpreted in terms of the physical processes that lead to the pattern. The sill is a measure of the overall variability of the process. The range is a measure of the spatial scale of the process. The nugget can be interpreted as the sum of two variances related to two distinct phenomena—the measurement error and the subgrid variability.

5.3.4 KRIGING WITH EXTERNAL DRIFT (KED)

KED (Deutsch and Journel, 1992; Goovaerts, 1997) uses additional or auxiliary variables to supplement the OK process. The auxiliary variables are used only if a clear-cut relationship exists between the variable to be interpolated and the auxiliary variable (or the variable being used as an external drift function). KED extends the covariance matrix of residuals to the auxiliary predictors while maintaining the overall similarity with the kriging process. KED can also estimate and sum the drift and residuals separately. Ahmad and de Marsily (1987) initially advocated this procedure. The technique was later renamed by Odeh et al. (1995) as Regression Kriging (RK). Goovaerts (1997) used the term kriging with varying local means for RK and Kriging with a trend model for the KED technique. Although similar, KED and RK differ in the computational steps used.

The theory is briefly described here. A spatial random function h(x) is considered, representing the water table elevation in an unconfined aquifer at location x. The empirical principle that the phreatic surface forms a subdued replica of the land surface can be mathematically expressed as follows:

$$h(x) = Z(x) + r(x) = Z(x) + d(x) \tag{5.1}$$

where $Z(x)$, the topographic elevation at point x, is viewed as a deterministic trend function and the residual, water table depth $d(x) = r(x)$, is a stationary spatial random function with nonzero mean and spatial covariance $\mathrm{Cd}(h) = \mathrm{Cov}[d(x); d(x+h)]$. The stationarity assumption for $d(x)$ is justified at spatial scales spanning several drainage basins. However, it would not be appropriate at the scale of a single catchment, where there is a trend in $d(x)$ from the groundwater divide down to the surface drainage network.

KED estimation of $h(x_0)$ from the water levels $h(x_i)$ in n surrounding wells requires the elevation $Z(x_0)$ at the point of estimation x_0 in addition to the topographic elevations $Z(x_i)$ at the n wells.

5.4 RESULTS AND DISCUSSIONS

The water table elevation data for the pre-monsoon season of 2016 was interpolated using the different interpolation techniques discussed in the previous section. While using the IDW interpolation, the weight parameter was set to $W = 0.1$, $W = 0.5$, $W = 1$, $W = 2$, and $W = 4$, and five resulting groundwater table elevation grids were obtained (Figure 5.4) corresponding to each weight value. Increasing the weight value gives less importance to distant points in the interpolation process.

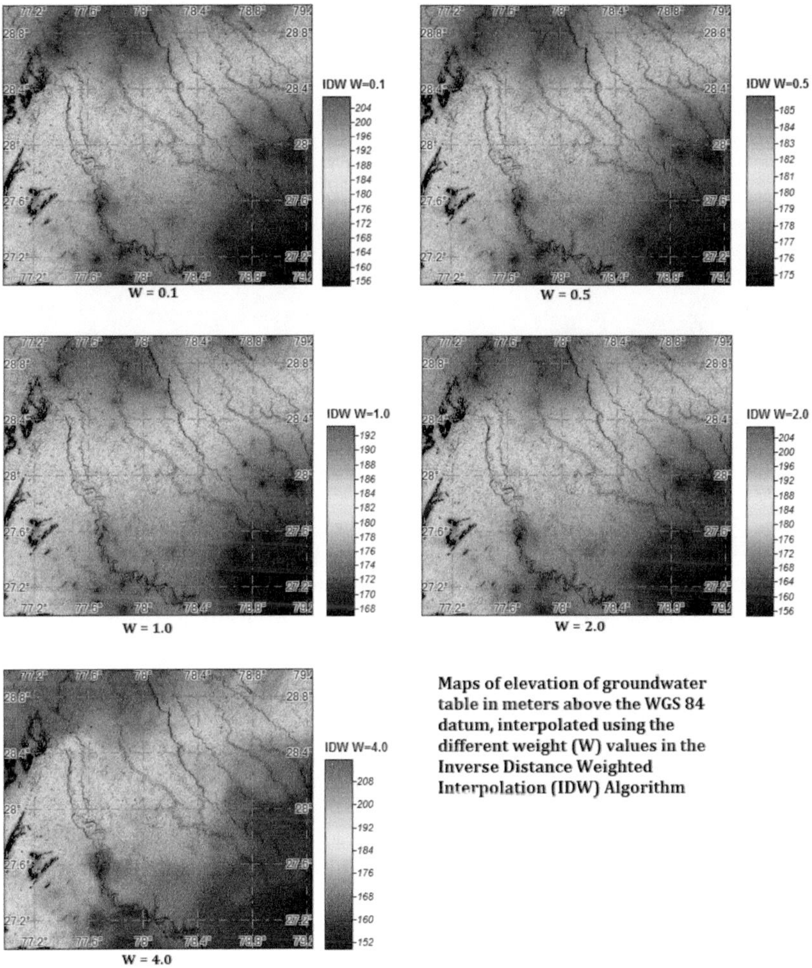

FIGURE 5.4 Maps of groundwater table elevation using inverse distance weighted interpolation.

In the Thin Plate Spline interpolation case, the regularization parameter was set to $R = 0.0001$ and $R = 1$, and two resulting groundwater elevation grids were obtained (Figure 5.5) corresponding to the two values of the regularization parameter. A higher value of the regularization parameter tends to smooth the final groundwater table grid, and a lower value tends to make a grid with relatively sharp breaks at the observation points. In this case, two extreme values were selected to assess this method's possible range of estimations.

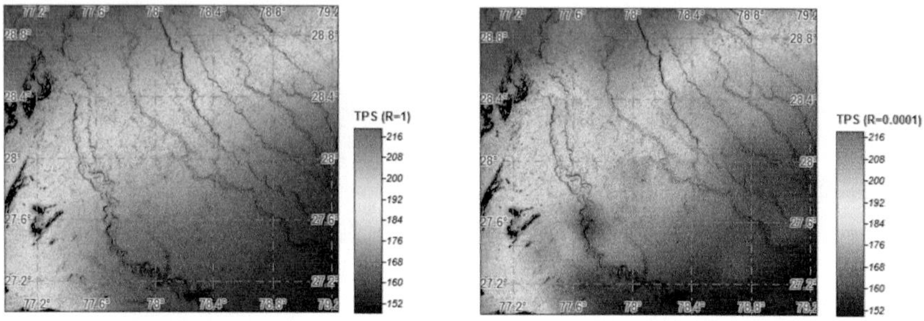

Maps of groundwater elevation in meters above the WGS 84 datum interpolated using the Thin Plates Spline (TPS) interpolation algorithm. The R values in the legend indicate the Value of the regularisation function used for interpolation.

FIGURE 5.5 Maps of groundwater table elevation using thin plate spline interpolation.

In the case of OK and KED, the variogram was initially estimated. The variogram shows nonstationary behavior, that is, the mean shows a spatial trend instead of stabilizing at a particular variance value (stationary behavior).

The observed variogram was fitted with a theoretical variogram (Figure 5.6) of the following form:

$$a + b \times x + c \times x^2 + d \times x^3$$

where $a = -23.423$, $b = 412.013$, $c = -438.213$, and $d = 253.78$.

The sill, range, and nugget parameters for the 150 groundwater elevations in the study area were obtained from the above variogram model and utilized in the interpolation process to solve the kriging weights.

Using the above variogram model, a groundwater elevation grid was obtained by OK interpolation as shown in Figure 5.7 (left image). The same variogram model is used for KED as the underlying dataset is the same. However, in the case of KED, a predictor grid is used to guide the interpolation process. The predictor grid is employed to estimate kriging weights beside the variogram model. The present study used a smoothed SRTM DEM of the study area as a predictor grid. The original 30-meter

SRTM DEM was filtered using a 50×50 smoothing kernel. Smoothing operation was undertaken because the groundwater table is only influenced by broad topographic features and not local topographic mounds and depressions. The groundwater elevation grid obtained using KED is shown in Figure 5.7 (right image). This grid has a greater reflection of the topographic structure in its spatial pattern, which is quite intuitive.

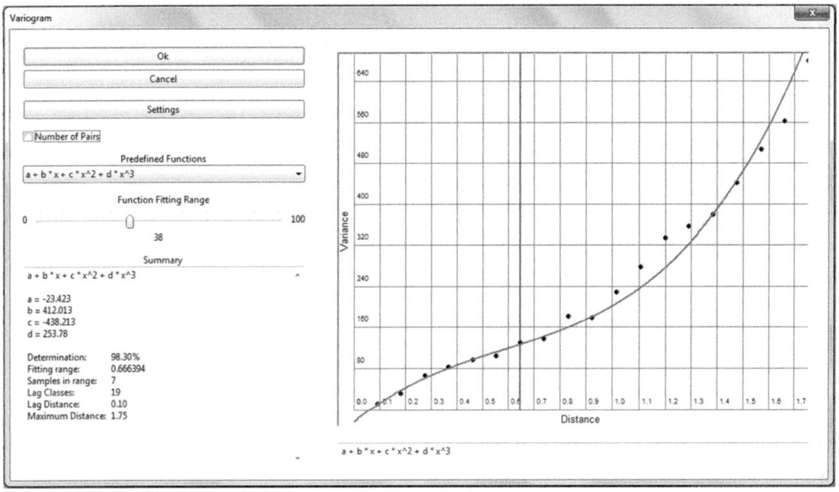

FIGURE 5.6 Experimental and theoretical variograms for the observed groundwater elevations in the study area.

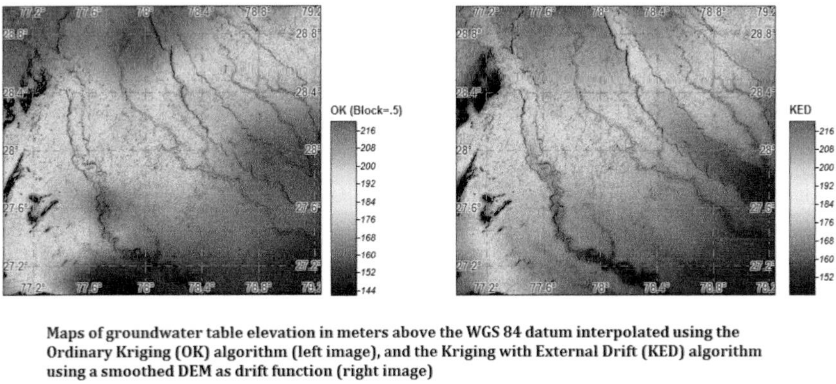

FIGURE 5.7 Maps of groundwater table elevation using Ordinary Kriging (OK) and Kriging with External Drift (KED)

A profile *X–Y* trending South West–North East was selected to assess the quality of interpolated groundwater table surfaces using different methods. Grid cell values along the profile points were extracted using all the grids interpolated with different procedures. The groundwater table elevation profiles were superimposed on the SRTM DEM (surface elevation) profile to visually compare the interpolated surfaces' reliability (Figure 5.8). From the profiles, it is clear that none of the IDW interpolations (Figure 5.8A) with varying weight parameters reliably reproduce the expected groundwater table. In the case of Thin Plate Spline interpolation (Figure 5.8B), a lower value of regularization seems to be reproducing the water table in a better way, although with many anomalies, for example, near the northeastern corner, the water table is going 8–9 m above the ground surface over the whole floodplain, which is clearly not possible. The same problem is observed in the water table obtained from OK (Figure 5.8C), which seems to perform marginally better than Thin Plate Spline.

Finally, the water table obtained from the KED method (Figure 5.8C) is observed to reproduce the variability in the most expected manner, that is, varying in accordance with the topographic surface. As per the theoretical understanding, the depth of the groundwater table is expected to be greater near the interfluves and lower in the floodplains. However, the groundwater table obtained from the KED model does not replicate this behavior. Hence, besides DEM, some other parameters, for example, Topographic Wetness Index (TWI) may also be used to overcome this problem.

5.5 CONCLUSIONS

Groundwater elevations in an unconfined alluvial aquifer were interpolated using different interpolation methods. It was found that the best interpolation method that replicates the original groundwater table to a high degree of precision is the one that uses auxiliary information to guide the interpolation process. In this study, the KED model using a smoothed DEM as a predictor performs the best. It is recommended that KED with smoothed DEM should be preferred as the interpolation method over other interpolation methods, especially for unconfined aquifers.

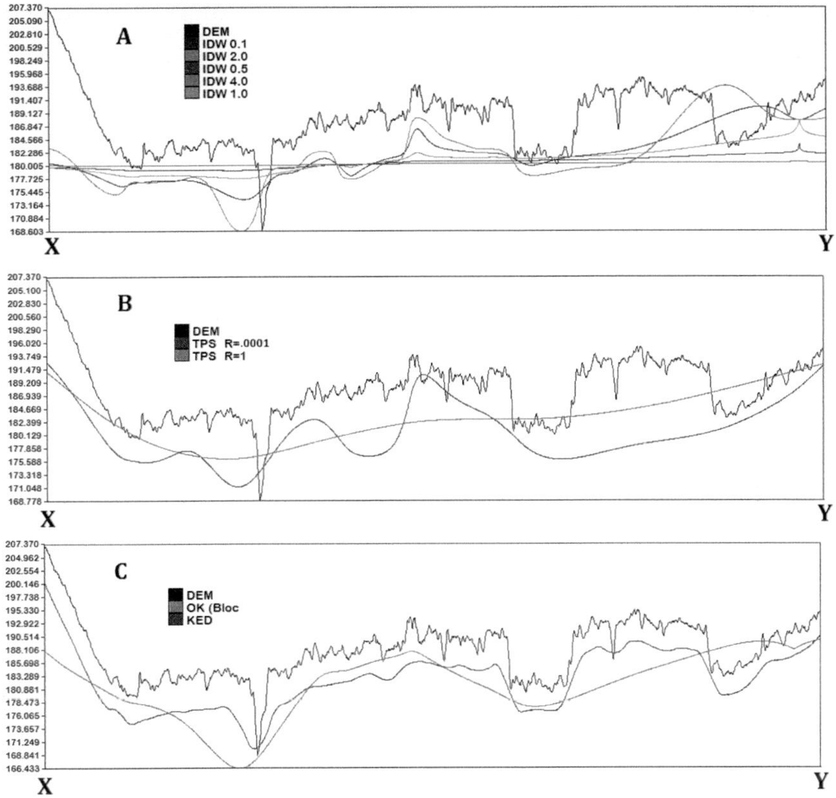

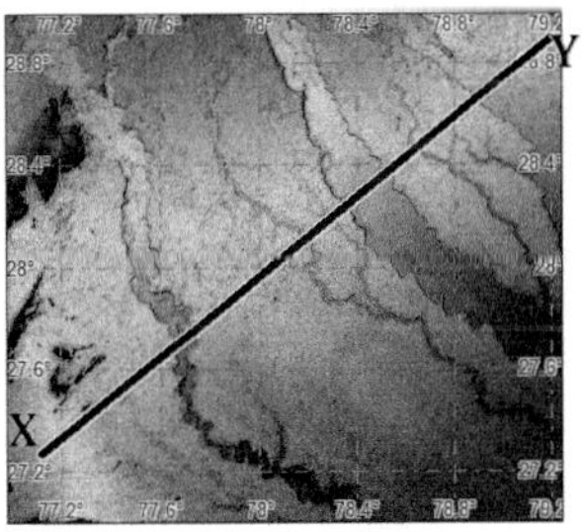

Topographic and groundwater table profiles along the line X-Y.

A. Groundwater table profiles interpolated using Inverse Distance Weighted Algorithm (IDW) with different weights.

B. Groundwater table profiles interpolated using Thin Plate Splines with varying Regularization (R)parameter.

C. Groundwater table profiles interpolated using Ordinary Kriging (OK), and Kriging with External Drift (KED) using a smoothed SRTM DEM as auxilliary variable.

FIGURE 5.8 Visual comparison of groundwater table profiles interpolated using different methods.

KEYWORDS

- **Digital Elevation Models**
- **Kriging with External Drift**
- **Inverse Distance Weighted**
- **Ordinary Kriging**

REFERENCES

Ahmad, S., and de Marsily, G. (1987). Comparison of geostatistical methods for estimating transmissivity using data on transmissivity and specific capacity. *Water Resources Research 23*(9), 1717–1737.

Bardossy, A., and Lehman, W. (1998). Spatial distribution of soil moisture in a small catchment. Part 1: geostatistical analysis. *Journal of Hydrology 206*, 1–15.

Bourennane, H., King, D., Chery, P., and Bruand, A. (1996). Improving the kriging of a soil variable using slope gradient as external drift. *European Journal of Soil Science 47*, 473–483.

Chiles, J. P. (1992). The use of external-drift kriging for designing a piezometric observation network. In: Bardossy, A. (Ed.). *Geostatictical methods: Recent Developments and Applications in Surface and Subsurface Hydrology. Proceedings of an International Workshop at Karlsruhe, Germany, July 17–19, 1990*, Paris: UNESCO, pp. 11–20 161 p.

Goovaerts, P. (1997). *Geostatistics for Natural Resources Evaluation*. New York: Oxford University Press.

Goovaerts, P. (2000). Geostatistical approaches for incorporating elevation into the spatial interpolation of rainfall. *Journal of Hydrology 228*, 113–129.

Hudson, G., and Wackernagel, H. (1994). Mapping temperature using kriging with external drift: theorgy and example from Scotland. *International Journal of Climatology 14*, 77–91.

Marechal, A. (1984). Kringing seismic data in the presence of faults. In: Verly, G., David, M., Journel, A.G., Marechal, A. (Eds.). *Geostatistics for Natural Resources Characterization. NATO ASI series C: Mathematical and Physical Sciences, Part 1*. Dordrecht, Netherlands: Reidel Publ, Vol. 122.

Odeh, I. O. A., McBratney, A. B., and Chittleborough, D. J. (1995). Further results on prediction of soil properties from terrain attributes: heterotopic cokriging and regression-kriging. *Geoderma 67*(3–4), 215–226.

CHAPTER 6

SPATIAL REGRESSION AND TEMPORAL ANALYSES USING GIS: A CASE STUDY OF INDIA

QAZI MAZHAR ALI[1], MOHAMMAD SAJID[2], RIZWAN AHMAD[2], ATEEBUR REHMAN SHERWANI[1], and AKRAM JAVED[3]

[1]*Department of Statistics and Operations Research, Aligarh Muslim University, Aligarh, Uttar Pradesh, India*

[2]*Interdisciplinary Department of Remote Sensing & GIS Applications, Aligarh Muslim University, Aligarh, Uttar Pradesh, India*

[3]*Department of Geology, Aligarh Muslim University, Aligarh, Uttar Pradesh, India*

ABSTRACT

Regression analysis is practical in modeling, examining, and exploring the spatial relationships among datasets. It helps explain factors responsible for the observed spatial patterns. This study explores the coal mining of the Singrauli area in central India from 1993 to 2020. The study uses the geographically weighted regression (GWR) method to model subsidence under the geographic information systems (GIS) that permit spatial variability of factors related to subsidence. This study identified four significant subsidence factors out of the seven analyzed: the coal level depth, surface slope, inclination, and thickness. Furthermore, the subsidence map is juxtaposed using the GWR method with these four factors as explanatory variables. Using the corroborated method, the

Sustainable Development: A Geospatial and Statistical Perspective. Qazi Mazhar Ali, Rizwan Ahmad, Irfan Ali, & Prasenjit Chatterjee (Eds.)

DOI: 10.1201/9781779643247-6

model evaluates and estimates the subsidence in the unmapped areas due to artificial landforms transformation. Remote sensing and GIS techniques are then applied for spatial–temporal analysis in detecting land use topography changes. Additionally, this study uses ArcGIS tools to analyze water level, area, shape, and volume change detection. A surface is generated via connection points located along the divide, while the ASTGTM-DEM represents the inner surface. The difference between the topography and the cap elevation represents the volume. By studying this decrease/increase trend behavior, one can derive from the regression analysis a third-degree polynomial, which indicates that the derived equation is representative.

6.1 INTRODUCTION

It is a fact that subsidence depends on time distortion produced by the overburden and readjustment of the ground's surface due to minerals/energy extraction. Subsidence related to mining can damage surface setups, constructions, farmlands, and engineering structures. It may also disturb the balanced underground water levels and the surfaces of the natural and human-made drainage systems. Deformation parameters can explain the subsidence area shapes. These parameters include, among others, horizontal and vertical displacements, curvature radius, tilt, and strain. They can also provide quantitative information related to likely hazards caused by surface deformation (Blachowski et al., 2014).

The mining subsidence nature has influenced several factors: mining depth, mined deposits thickness and inclination, manufacturing methods, tectonic movement, local geological situations, operating speed (underground), and mining period (Kratzsch, 1983; Knothe, 1984; Popiołek, 2009). Several researchers have conducted many investigations regarding subsidence in various countries via observations with the hope of controlling the adverse effects of underground coal mining in the late 1990s (Kuscu, 1991). Several studies in the past attempted to develop predicting models for land subsidence and its related hazardous damages.

Based on the literature, mining subsidence investigations are divided broadly into subsidence monitoring and observation and subsidence prediction and related hazards. The present research focuses on subsidence monitoring on the mining grounds via a geodetic approach (GPS, precise leveling) with the help of modern techniques like remote sensing. Several workers have also used RS and GIS to monitor and assess coal mining's impact in Singrauli coalfield (Khan and Javed, 2012).

Current studies aiming at investigating damages hazards related to subsidence mining include Oh and Lee (2011), where they used "frequency ratio, weights of evidence, logistic regression, and artificial neural network (ANN)." The study aimed to detect and quantify the association between some factors, such as draft depth, slope, drift distance, groundwater level, geology, permeability, and land use, and the location of land subsidence for realizing the hazardous ground subsidence produced as a result of coal mines abandoned in Korea. Thus, the relations were utilized as factor ratings in GIS overlay analysis to create ground subsidence hazardous indices and maps.

GIS-based spatial analysis has been conducted in different areas and fields with varying decision-making dimensions (de Moura and Procopiuck, 2020; Rodríguez-Espinosa et al., 2020; Unwin, 2020; Banerjee et al., 2020; Fischer et al., 2019; Verhagen, 2018). Moreover, hazards relating to land subsidence have been investigated in Korea using the decision tree approach and GIS incorporating significant factors (Lee and Park, 2013; Park et al., 2014). They produced and verified hazard maps for the study area based on the land subsidence. A study based on ground deformation phenomena caused by underground salt deposit exploitation that affects a town in Bosnia Herzegovina has been reported by Stecchi et al. (2012). A multicriteria analysis approach has been proposed to evaluate the susceptibility of an area to subsidence damage. Thus, making use of an external application is based on the GIS integration approach. GIS analytical functions have been used to assess geotechnical risks related to the underground coal mines abandoned in South Africa (Mathey, 2013).

The "classification and regression tree theory" has been applied in testing a mining site to evaluate the damages produced by surface deformation to concrete buildings in Poland (Malinowska, 2014). Among many assumptions underlying the ordinary basic regression model is the independence of observations. The premise may not be valid always for data related to spatial units, as according to Tobler (1970), "Everything is related to everything else, but near things are more related than distant things." There is some spatial dependency in the variables because nearby locations tend to have more similar values. Also, the model's residuals may exhibit similar dependence. The spatial dependency will be pronounced if the plotted neighboring units have a common sign and magnitude for the bare regression residuals on a map. Due to this behavior of spatial data, the parameter estimates for the basic model are affected. Hence, the parameter estimates could be inefficient due to the spatial structure of the model residuals, which return a more significant standard error of the parameters. Making inferences based on such estimates could have considerable implications since they are biased and inefficient. Data spatial structure happens when a unit's dependent variable value is influenced by a nearby unit's independent variable value leading to biases in the estimates (Charlton and Fotheringham, 2010).

The present study proposed a spatial regression methodology based on GIS weighted in investigating related land subsidence in the mining of Singrauli coal mines areas in India's central state. Blachowski and Milczarek (2014), in an initial study concerning the site, found that the conventional subsidence mining methods based on empirical assumptions like the Knothe theory (Knothe, 1984) yielded unreliable results for thin, abundant steeply inclined coal seams.

Brundson et al. (1996) and Fotheringham et al. (2002) propose a geographically weighted regression (GWR) as a method that permits one to know the manner of spatial-specific area variations for a given phenomenon (e.g., subsidence). Thus, GWR can be regarded as a multiple linear regression's extension spatially, as it considers the local parameters. Hence, the method allowing mining subsidence factors spatial variability should undoubtedly give better mining subsidence estimates than conventional regression methods such as the "ordinary least squares" (OLS) or the above-mentioned empirical procedures for those mining conditions that are complex.

Subsidence can be predicted using GWR in areas too limited for mapping or where the surface displacement measures are unavailable. Recently, the method has been helpful in socioeconomic and environmental issues with spatial variability, and therefore, is gaining sufficient attention. The next section presents a detailed description of GWR applications and their underlying theory. The present study uses the GWR method to develop a model based on spatial regression for analyzing associations between subsidence which is the response variable in the model. Other standard subsidence mining factors are taken as explanatory variables. Thus, the factors analyzed for mining subsidence were the thickness and coal panels' site, inclination and coal panels' depth underground level, slope and elevation (surface topography), coal parcels' distance, and mining system used therein. This study aims at achieving the following objectives:

1. To predict mining subsidence using the GWR model,
2. To identify factors that are statistically significant for mining subsidence, and
3. To develop spatial regression models in part of the Singrauli coal mines for mining subsidence.

6.2 STUDY AREA, GEOLOGY, AND TOPOGRAPHY

The study area is located at the geographical center of India, between the Singrauli district and Sonebhadra district of Madhya Pradesh and Uttar Pradesh states, respectively. It is resting on the latitudes 24°00′ to 24°15′N and longitudes 82°30′ to 82°45′E geo-coordinates (Figure 6.1). The coalfield's landmark is 2201 km^2 in total. However, the area exploited for coal is only 300 km^2. It is an excellent superthermal power plant location because of the power-grade coal reserves and the nearby water reservoirs availability (Govind Ballabh Pant Sagar). Aluminum plants and cement industries can also be suitable for the location.

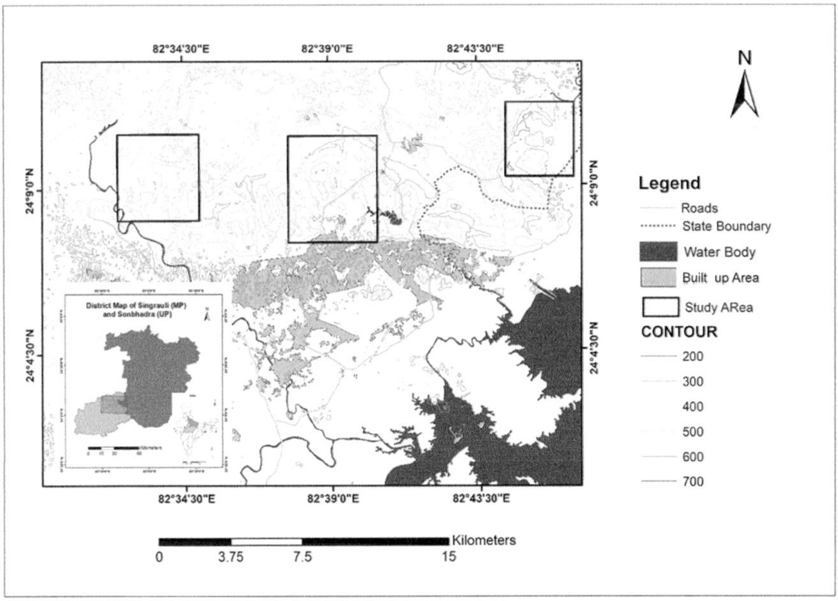

FIGURE 6.1 Study area location and the active mining grounds' boundaries.

There are 1789.41 million tons of coal reserves in Singrauli coalfield. Also, there is a steep rise in coal production from 30.70 million tons in 1993 to 67.67 million tons in 2010. According to Jamal et al. (1991), the area is expected to produce 25% of India's thermal power need because of its thermal power generation potentials. The district Singrauli is a dry tropical monsoonal from November to June, while the atmosphere is very humid during the rainy season. The region's average rainfall is 1177.06 mm for the last 26 years, while the minimum and maximum are 798.60 and 1457.30 mm, respectively, in 1978 and 1979.

The region characterizes by a series formation as a geological sequence recognized on lithic broadly characterized within the Gondwana rocks of the Singrauli coalfield. They are "Talchir, Barakar, Barren Measure, Raniganj, Precambrian, and Mahadeva" in ascending order. However, not all of them have been reported from the study area; only four "Precambrian, Talchir, Barakar, and Barren Measure" were reported. Barakar formation is the most common geological formation covering 515.18 km^2 in the area, followed by "Talchir" covering 118.94 km^2, then Precambrian and Barren measures covering 57.63 km^2 and 16.89 km^2, respectively. The northern part of the area is occupied by structural hills formed of resistant Precambrian rocks with an elevation between 270 m and 620 m.

The study has been carried out using both primary and secondary data sources. The present study has utilized remote sensing data of Landsat 8 data of April 29, 2020 (Figure 6.2). The plateau is composed of coal-bearing Gondwana rocks structurally. To the central area are low-lying flats with gentle-undulating topography suitable for primary agricultural practices (Majumder and Sarkar, 1994). The main dendritic perennial streams are the "Kachin, Mayar, Motwani, and Baliya Nala" transverse via the "Singrauli coalfield" seasonally. The developmental expansion of this area began with two dams constructed in the 1950s on the Rihand River. The primary purpose of the dam's construction as reservoirs was mainly for irrigation. However, it generates 400 MW of hydropower. The area was preoccupied densely with tropical deciduous forests before the advent of mining activities. The coal mining intensity in the region has dramatically redefined the environment with adverse environmental effects due to industrialization. The land cover, quality of air, and water in the study zone are greatly affected.

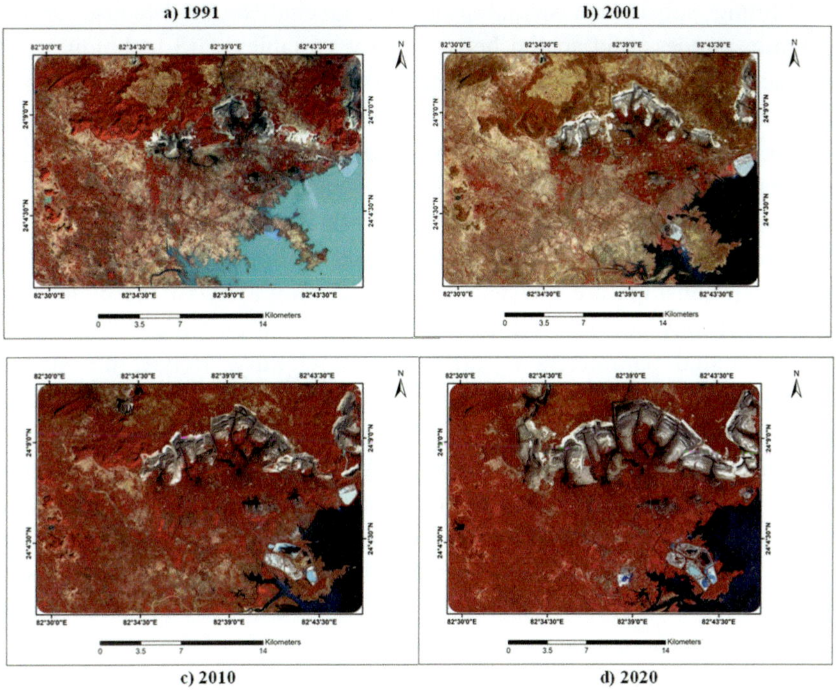

FIGURE 6.2 Temporal satellite data of Landsat used for Singrauli coal mine pit temporal analysis: (a) 1991; (b) 2001; (c) 2010; (d) 2020.

6.3 METHODOLOGY—*SPATIAL REGRESSION ANALYSIS*

The concept of "spatial regression" has been thoroughly dealt with and documented (Brundson et al., 1996; Fotheringham et al., 2002). The application of the idea has been discussed by Charlton and Fotheringham (2010) and Charlton and Fotheringham (2009), respectively. There are comparatively limited studies associated with land subsidence analysis and possibly none relating to mining-induced subsidence using this method. Shang et al. (2011) use the GWR method in Taiwan to explore the spatiotemporal variations using the data from an observed groundwater level, benchmark elevations, and examination of the spatial heterogeneity. The results revealed that the groundwater pumping activity relates to the occurrence and formation of land subsidence.

Mining subsidence spatial modeling is challenging because of the complex process and spatial heterogeneity variables. GIS techniques can handle the voluminous datasets required for modeling subsidence, especially the spatial data processing method. Hence, this method is applicable in developing subsidence models and related analyses. For instance, the correlation between subsidence as a variable and other factors (one or more independent variables) can be determined using regression analysis.

The "Spatial regression" methods could work on the assumption that these relationships are consistent with the geography of the concerned area, or it may consider the features spatial locations, thereby allowing the estimated parameters to differ in the vicinity. Hence, the spatial variability in the relationships of the response and explanatory variables can be reflected along with the "OLS" method. The OLS method takes into consideration that relationships are geographically stable. Moreover, the GWR permits exploration and spatial distribution analysis of variables by fitting a regression model to each point in space, thereby studying spatial variability and relationships. The GWR helps in providing a local model for the predicted scenario about the response variables and identifies the variables' coefficients at each location.

Charlton and Fotheringham (2009) have defined the generalized model for multiple linear regression shown in Equation 6.1 as

$$y_i = \beta_0 + \beta_1 x_{1i} + \beta_2 x_{2i} + ... + \beta_m x_{mi} + \varepsilon_i, \forall i, i = 1, 2, ..., n \tag{6.1}$$

where y_i denotes the observed value of the response variable calculated at the ith location x_i is predicted observations relating to explanatory variables, β_j, $j = 1,2,\ldots, m$ are the model coefficients or parameters, and ε_i is the error term.

The parameters β_j, $j = 1,2,\ldots, m$ can be estimated as

$$\hat{\beta} = (X^T X)^{-1} X^T y \tag{6.2}$$

where X represents the matrix of independent variables values $(X^T X)^{-1}$ and is the inverse of variance and covariance matrix, whereas y is the observed values vector. If the data have different kinds of uncertainty, weighting the observations in the regression model can improve the accuracy of the model. Hence, Equation 6.2 with weighting the observations in the leading diagonal of a Square matrix W as follows:

$$\hat{\beta} = (X^T W X)^{-1} X^T W y \tag{6.3}$$

The above-discussed regression model can help measure the actual parameter statistics how far from its prediction. Whether variables used in the model contribute significantly to the model, the *t*-statistics used to verify the model results.

The general multiple linear regression model equation 6.1 with a spatial dataset which is some linear function of the locations (u_i, v_i) can be rewritten as follows:

$$y_i = \alpha_0 + \alpha_1 u_i + \alpha_2 v_i + \beta_0 x_{1i} + \beta_1 u_i x_{1i} + \beta_2 v_i x_{1i} + \gamma_0 x_{2i} + \gamma_1 u_i x_{2i} + \gamma_2 v_i x_{2i} \quad (6.4)$$

Then Equation 6.4 can be approximated to a simple three parameters and two independent variables, that is,

$$y_i = a_i + b_i x_{1i} + c_i x_{2i}$$

where

$$a_i = \alpha_0 + \alpha_1 u_i + \alpha_2 v_i$$
$$b_i = \beta_0 + \beta_1 u_i + \beta_2 v_i$$
$$c_i = \gamma_0 + \gamma_1 u_i + \gamma_2 v_i$$

The disadvantage of the model is defined in Equation 6.4 is that before modeling the situation, the analyst must determine the expansion nature of the parameters. The spatially varying parameter estimates can be easily computed and mapped if the observations' locations are known. Equation 6.4 was not considered critical underlying assumptions in the ordinarily defined regression models. The basic assumption of the regression model is defined in Equation 6.1. Observations must be independent, and they may not always be accurate, specifically with spatial datasets. Since nearby locations, dataset observations may have similar values. Hence, the study variables in the model exhibit spatial dependence, not only in the study variables but also the model's residuals might be spatial dependent.

The residuals in similar spatial dataset units have a similar magnitude and sign based on observations. Suppose the observations have the spatial structure in the residuals from the model. In that case, it leads to inefficient estimates of the study parameters, and hence it can be concluded that the standard errors of the parameters are significant. A spatial structure datasets the value of one dependent variable in one spatial observation that may be affected by the independent nearby observations' dataset, leading to misguiding the analyst about parameter estimations. Hence, parameter estimates are both biased and inefficient. Spatial heterogeneity is another vital issue in spatial dataset modeling. Wherein parameters of the models are themselves a function of location. For the heterogeneous spatial data modeling, the GWR is more appropriate. In GWR, the study parameters can be estimated for any random site conditionally. A dependent variable and a set of independent variables have been measured for a known location point.

Considering Tobler's observation regarding datasets neighboring and similarity to estimate the model parameters at some location u^2, the neighboring observations have more weight than other observations in the estimation. The mathematical expression for the GWR of Equation 6.1 is as follows:

$$y_i = \beta_{0i}(u) + \beta_{1i}(u)x_{1i} + \beta_{2i}(u)x_{2i} + \cdots + \beta_{mi}(u)x_{mi} \tag{6.5}$$

The symbol $\beta_{0i}(u)$ is used to describe a relationship for a specified location u in the study area. A response variable can be predicted if the measurement for the explanatory variables is available at the location u. Equation 6.3 considering the weights for location u with the other observations in the datasets relatively, the estimator is as follows:

$$\hat{\beta}(u) = (X^T W(u) X)^{-1} X^T W(u) y \tag{6.6}$$

where $W(u)$ is a square matrix of weight relative to the location u, $X^T W(u) X$ is the geographically weighted variance–covariance matrix, and y is the dependent variable values vector. The $W(u)$ square matrix with geographical weights in leading diagonal and whereas zero in off-diagonal

$$W(u) = \begin{bmatrix} w_1(u) & 0 & \ldots & 0 \\ 0 & w_2(u) & \ldots & 0 \\ \vdots & \ddots & \ldots & \vdots \\ 0 \ldots & w_i(u) & \cdots & 0 \\ \vdots & \ddots & \ldots & \vdots \\ 0 \ldots & 0 & \ldots & w_n(u) \end{bmatrix}$$

The kernel uses to compute the weights $w_i(u)$. The Gaussian shape kernel is defined as follows:

$$w_i(u) = e^{-\frac{1}{2}(d_i(u)/h)^2}$$

where $w_i(u)$ denote the ith geographical weight observation relative to the location **u** $d_i(u)$ in the dataset. It is some distance measure between the ith observation and the location **u**, and bandwidth h. The bandwidth (h) is defined in the same units as the coordinates in the dataset. If the bandwidth gets larger, the weights approach to unity and GWR regression model 6.5 approaches global OLS 6.1. The fitted model can be influenced by choice of bandwidth (h) and the kernel shape. If the sample observations are regularly spaced in the study area, then a kernel with fixed bandwidth is good for modeling. Suppose the sample observations are not regularly spaced but are clustered in the study area. In that case, a kernel allows accommodating such irregularities by way of its size increment when sample observations are less and conversely reducing its size when dense sample observations are observed. The specified kernel can calculate the weight [$w_i(u)$]. Here, the value of any observation having a distance more than its bandwidth will equate to zero and be excluded from the calibration locally.

The model goodness of fit is measured using the r^2 or adjusted r^2 values. In GWR, the model's adequate number of parameters is considered when computing the goodness of fit measure. Preferably, if several models are compared, the adjusted r^2 value can compensate for the number of variables or parameters in the fitted model. The adequate model parameters numbers depend on the independent variable's numbers and the bandwidth. The corrected Akaike information criterion (AIC$_c$) proposed by Hurvich and Tsai (1993), is best to measure the GWR model's goodness of fit. That is,

$$AIC_c = 2n\log_e(\hat{\sigma}) + n\log_e(2\pi) + n\left(\frac{n+tr(S)}{n-2-tr(S)}\right) \tag{6.7}$$

In Equation 6.7, the observations are represented by $\boldsymbol{n}$ and $\hat{\sigma}$ denote the residuals' standard deviation estimates and $\boldsymbol{tr}$ ($\boldsymbol{S}$) are the hat matrix's trace $\boldsymbol{S}$. The tr ($\boldsymbol{S}$) yields the number of model parameters, the trace sums the leading diagonal values of $\boldsymbol{S}$.

The AIC$_c$ measures the quality of the statistical model relative to each other. Because whenever data are plotted on a particular distribution, some information losses due to error; therefore, AIC try to estimate the relative amount of loss information from its actual model and select the model with a minimum loss as the best model. The typical information distance is referred to as Kullback–Leibler information distance; if the value AIC$_c$ = 3, the two models under comparison are said to be equivalent.

At the regression points, the GWR is provided estimates relating to all parameters and standard errors associated with them. Suppose the regression points are the same as the sample points; then, the GWR yields estimates for residuals and standardized residuals for the dependent variable.

The present study uses a weighted spatial regression (WSR) method for variable hypothesis testing that is likely to represent the computed subsidence about response variable during 30 years (1993–2020) of mining in the Singrauli coal mine area the time series analysis of satellite observations. The analyzed subsidence factors relating to independent or explanatory variables take into consideration the following factors, namely, "site and thickness of coal panels, depth of coal panels under the ground, the slope of coal panels, surface height, surface angle, and distance of mining system from mined coal panels."

The following steps are considered in the research methodology:

1. *Raster preparation for mining subsidence factors spatial representations*—response and explanatory variables for spatial regression analysis.
2. *Exploratory analysis.* This involves general analysis of spatial regression models and identifying and selecting variables that provided the best explanation for the response variable (subsidence).
3. *Constructing and analyzing the GWR model* with the same set of variables.
4. *Subsidence prediction map* construction with the GWR model.

6.4 RESULTS AND DISCUSSION

The current study area is the Singrauli coal mine, one of India's central state's major functional coal mines. The land subsidence has been calculated for the analyzed part using the methodology presented in Section 6.3. The subsidence map construction has been done by taking the "digital elevation model" (DEM) created from ASTER DEM (1:10,000 scale of the surface's DEM with 1.5 m vertical and 20 m horizontal resolution, imported by USGS datasets). The results show that the maximum subsidence in the present study area for 30 years is 27.9 m, while the elevation is +48.7 m with ±3.0 m accuracy. The anthropogenic forms developed over time, such as mining waste dumps for the terrain, bear their association with the areas that have been formed later.

Moreover, "coal parcels and coal seams" from the vector data signifying as a 3D (XYZ) Polygons stored in the ArcGIS database with its attributive information (Figure 6.3). The construction of the maps on mining subsidence factors has been done. The Singrauli mine plans have been computed at scales 1:2000 and 1:5000 for nearly 30 years of extracting coal, and the database has been created by coding the mining maps to a vector format (mesh). The explanatory variables that were analyzed, thus, included: (1) the coal panels width, (2) the coal panels underground depth, (3) the coal panels' inclination, (4) the site and distance from the coal panels, (5) the surface digital elevation model, and (6) slope of surface used (Table 6.1). The present study has prepared the raster maps with 100 m pixel size by dividing the study area into 100 m × 100 m polygon mesh for all the possible explanatory variables spatial representations. The use of raster datasets (response and explanatory variables) has been considered in extracting a specific variable value and allocating these variables to the mesh spatially convergent polygons used in spatial regression analysis.

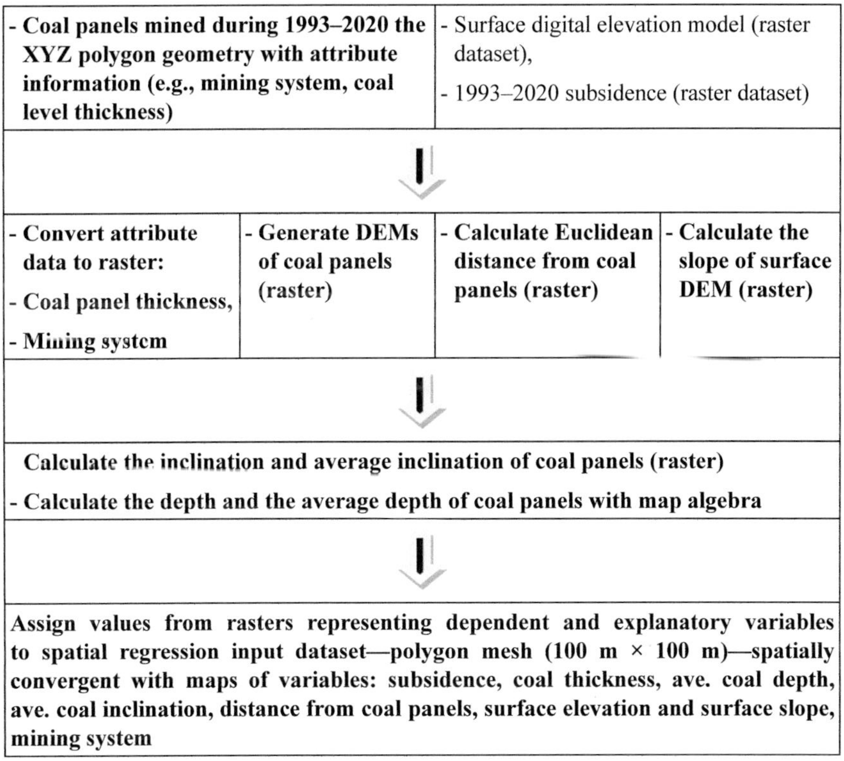

FIGURE 6.3 Workflow of data preparation and preprocessing and source of spatial data in the Singrauli mine pit database.

The data flowchart preparation for spatial regression is depicted in Figure 6.3. For the dependent (subsidence) and explanatory variables, the final thematic maps have been shown in Figure 6.4, and Table 6.1 presents the explanatory variables. For the current study, ArcGIS software version 10.2 has been used.

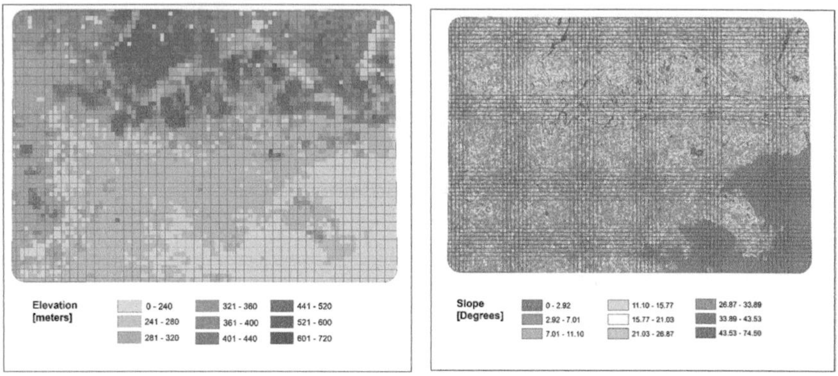

FIGURE 6.4 Spatial representation of the regression models' variables.

Because of the coal extraction complications, this study assumed the following simplifications:

1. The underground coal panels' depth was calculated as a difference between the surface DEM and the 20 mined coal seams' mean elevation (Figure 6.4).
2. The coal panels inclination has been computed as the average of all the coal panels inclination used.

In the analyzed part of the study area, the range for the thickness of a particular coal level has been found between 2.7 and 6.63 m, while the total thickness of all coal levels combined varies from 2.7 to 19.97 m. Moreover, the average inclination of coal panels is between 2° and 3°, whereas the change in the coal panels mean depth ranges from below 80 m to approximately over 400 m under the ground. The extraction process for the coal was done predominantly by using a long-wall terrace. The mining system variable has not been incorporated in the final regression model, as it is statistically insignificant.

The present study divides the area under consideration into two parts: one representing the calculated subsidence region, that is, 80%, and the

TABLE 6.1 Coal Seam Explanatory Variables (Sample Size 100 m × 100 m).

Coal Seam Variables (Sample Size 100 m ×100 m)									
Variables	**Classes/Range**								
Elevation (m)	160–240	241–280	281–320	321–360	361–400	401–440	441–520	521–600	601–720
Slope (Percent rise)	0–7.07	7.07–18.38	18.38–29.70	29.70–42.43	42.43–57.99	57.99–76.37	76.37–04.66	104.66–56.99	156.99–360.67
Aspect (Direction)	North (0–22.5)	Northeast (22.5–67.5)	East (67.5–112.5)	Southeast (112.5–157.5)	South (157.5–02.5)	Southwest (202.5–247.5)	West (247.5–92.5)	Northwest (292.5–337.5)	North (337.5–360)
Subsidence (m)	–159 to –80	–79 to –40	–39 to 0	1–40	41–80	81–120	121–200	201–320	320–350
Inclination (degree)	2–3	2–3	2–3	2–3	2–3	2–3	2–3	2–3	2–3
Depth (m)	80	120	160	240	280	320	340	360	380
Effective Thickness (m)	19.97	17.74	11.42	10.43	6.63	5.68	4.48	3.29	2.7

other signifying the elevation area calculated between 1993 and 2020 found to be 20% of the total area. During the period that has been analyzed, the latter part has also experienced mining-related subsidence. Still, it was not detected due to the growth of anthropogenic landforms, for example, waste dumps. The subsidence area has been used in constructing and validating the WSR model describing the mining subsidence basin. Figure 6.5 shows the extent of both zones. It has been found that the location of calculated subsidence comprised 433 square polygons, while 638 irregularly grouped square polygons constitute the prediction area.

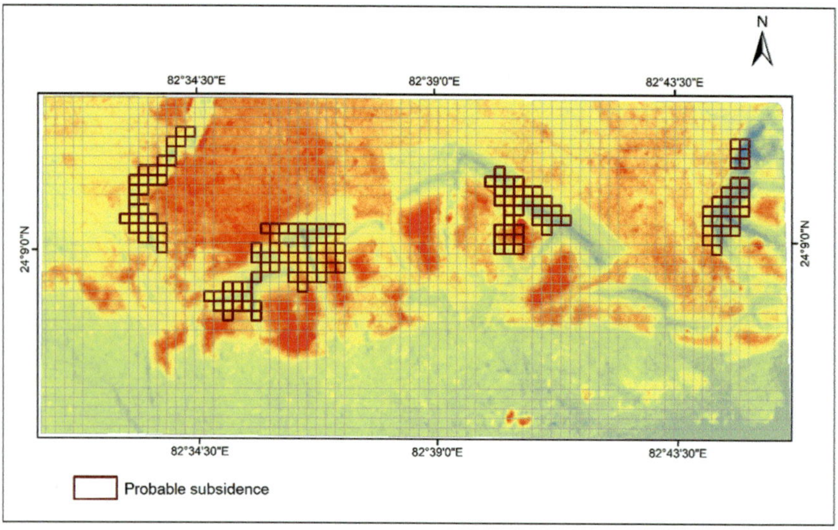

FIGURE 6.5 Hybrid map of possible subsidence/subsidence predicted for the elevated area.

6.4.1 THE GWR RESULTS

The coefficient of individual features in the space is calculated using the GWR model; these may vary, however, the local changes are captured. The mapping of the resulting coefficient estimates has been done to study regional variations in the parameter discussed. Generally speaking, every feature estimate consists of a unique weight matrix that characterizes the significance of individual observed data in calculating parameters at each site. It is noticed that the closer to each location an observation is, the more significant it shows the weight. For GWR model fitting, the AIC_c was used as a tool for diagnostic purposes extensively.

The GWR results showed a considerably more significant value of the R^2 statistic equal to 0.78 as compared to the value of 0.54 attained in the OLS analysis method. For the GWR model, it was found that the AIC_c value was 17,280.62. In the OLS analysis, 22,732.46 greater R^2 values and lesser than AIC_c values, showing that the GWR model is much superior to the conventional regression model and is, thus, successful in capturing spatial heterogeneity for explanatory variables. The spatial autocorrelation test has been run on the GWR model, the residuals with the Moran's, indicating a random array having a *Z*-score of 1.23022 and a *p*-value of 0.15487. It should be noted that the *Z*-score is the standard deviation, while the *p*-value is probability. Table 6.1 presents the independent variables. The calculated surface on the maps changes between 1993 and 2020 positively due to new landforms development, as shown by the blank areas. Eventually, the GWR model became useful in estimating subsidence in regions with positive topographic forms newly formed due to industrial growth, for example, overburden, thereby elevating the ground level. In the case of the GWR model, it has been predicted that the maximum subsidence for these parts is –11.5 m, with –4.3 m as the mean subsidence value.

As has been discussed by Choi et al. (2010), the results thus obtained in the study of the Singrauli mine are in conformity with other subsidence analyses in previous coal mining areas about the factors related to the depth and surface slope. It may, thus, include a fuzzy approach to ground subsidence mapping and several methods involving ANNs and statistical analysis (Lee and Park, 2013; Oh and Lee, 2011). The focus of these studies is predominately on cases related to the horizontal, single coal seam. In contrast, none of the studies has considered the factors associated with thickness, exploited coal panels' inclination, and the mining system. For elucidating the character of subsidence mining in this study, concerning multiseam deposits with a variety of thickness and inclination of the coal panels mines, it is seen that the first two of these factors are significant. As considered in the literature, these features are tough to represent because of their complex nature and, therefore, are known to affect subsidence mining (Kratzsch, 1983; Knothe, 1984; Sinclair Knight Merz Pty Ltd, 2014). This study observed that the subsidence having higher values are produced on the surface in case of greater aggregated "thickness and steeper inclination of coal panels" during the Singrauli study. Moreover, higher subsidence values have been associated with shallow exploitation under valleys and gently sloping areas.

Consequently, a mining subsidence map from 1993 to 2020, post-mining period to the culmination of mining activity, has been prepared for the Singrauli mine area.

It is a hybridized map designed for the subsidence calculated from the estimated cartographic data using the validated GWR model considering the four explanatory variables identified in this study. The map is presented finally in Figure 6.5, showing an undistorted subsidence basin picture due to concurrent land transmission. It was developed due to the entirety mining of the area, including concealed parts due to new landform development, which provides subsidence damage hazard information. The darker-blue portion of the map is the most significant subsidence, reflecting the mining parameters' complexity such as "thickness and inclination" and those with multiple coal seams mining intensity. These areas are primarily sites having concurrent waste dump spatially. The zones also have other forms of anthropogenic terrain representation.

The proposed approach enhances modeling on subsidence mining and secondary subsidence for harsh conditions with complexity. The present study results show that the WSR model is more reliable for subsidence mining. Therefore, the method can serve as an alternative to conventional methods, extending the subsidence modeling approach to predict and assess the process for areas with inadequate survey data and related cases to complicated mining situations.

6.4.2 TEMPORAL ANALYSIS

The temporal analysis is based on 1991, 2001, 2010, and 2020 satellite data analysis. It is found that during 1991–2001 an area of 1.07% (Table 6.2) has seen a change in land use/cover category, whereas, during 2001–2010, 2.32% (Figure 6.6) area reported a change in land cover/use category and during 2010–2020, 2.82% area reported change in land use/land cover category. The overall change during the 1991–2020 period is 6.21%, which suggests the area's dynamic land use, where the area has changed from one land use/cover to another land use/land cover (LU/LC) category. This analysis suggests that the area has seen rapid changes during the last 27 years due to industrialization and the expansion of coal mining activities. The coal mining in the study area has a significant impact on LU/LC, as the analysis suggests that mine pits, overburden dumps, ash ponds have seen a significant increase in their areas from 1991 to 2020, namely, 7.65 km^2, 59.11 km^2, and 11.55 km^2, respectively. Coal mining has also affected dense forests, which has reduced by about 51.78 km^2 during 1991–2020 since the expansion of coal mining has taken place in virgin areas.

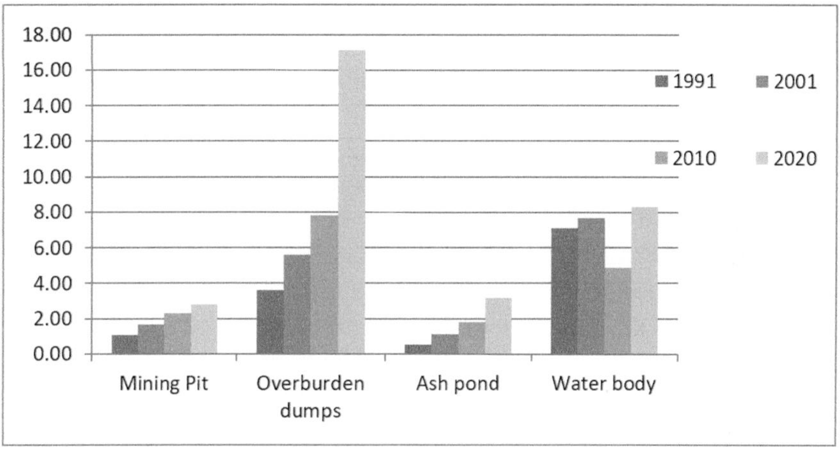

FIGURE 6.6 Temporal variations in the region and factors affecting land subsidence.

TABLE 6.2 Land Use Classes Affecting Land Subsidence in the Study Area

Land Subsidence Driven Factors/Classes	1991	2001	2010	2020
	Area in (%)	Area in (%)	Area in (%)	Area in (%)
Mining pit	1.07	1.66	2.32	2.82
Overburden dumps	3.61	5.60	7.80	17.11
Ash pond	0.53	1.14	1.82	3.16
Water body	7.11	7.70	4.87	8.32

6.5 CONCLUSION

For the present study, a multifactorial WSR approach based on GIS has been used for modeling and analyzing subsidence mining and its associated hazards in coal mining areas characterized by complicated mining situations. For the Singrauli coal mine located in Madhya Pradesh, the central state of India, the methodology used here aims to create a spatial subsidence model and predict subsidence mining for areas that do not cover by deformation surveys.

The research focuses on seven common factors for mining subsidence. Four are statistically significant: thickness, inclination, coal panels depth, and surface slope. However, the last two factors correspond with outcomes presented in other studies of mining subsidence that consider more simple conditions that may include horizontal mining and single coal seam mining methods. The first two factors have not been considered in most other published studies and are necessary for multiseam "coal mining and steeply inclined coal panels" as in the Singrauli mining area. Temporal analysis suggests that the area has seen rapid changes during the last 27 years as a result of industrialization and expansion of coal mining activities.

The GWR was used in estimating parameter and associated regression points standard errors. GWR can predict the response variables, residuals, and standardized residuals if the sample points are the same as the regression points. As a result, a hybridized subsidence map can be calculated from cartographic data processing. Using four identified statistically significant explanatory variables, this study used the validated GWR model to estimate mining subsidence.

The present research developed a general mining subsidence hybrid map for the Singrauli study area. The proposed study methodology can be extended and used to model mining subsidence for regions under the mining influence. Moreover, the procedure is also relevant for mining conditions that are complex and complicated and can thus be applied as an alternative for other methods.

KEYWORDS

- **modeling**
- **spatial regression**
- **prediction**
- **GIS**
- **time series**
- **temporal change**

REFERENCES

Banerjee, S., Kabir, M. M., Khadem, N. K., and Chavis, C. (2020). Optimal locations for bikeshare stations: A new GIS based spatial approach. *Transportation Research Interdisciplinary Perspectives*, *4*, 100101.

Blachowski, J., and Milczarek, W. (2014). Analysis of surface changes in the Wałbrzych hard coal mining grounds (SW Poland) between 1886 and 2009. *Geological Quarterly 58*(2), 353–367. https://doi.org/10.7306/gq.1162

Blachowski, J., Milczarek, W., and Stefaniak, P. (2014). Deformation information system for facilitating studies of mining-ground deformations, development, and applications. *Natural Hazards and Earth System Sciences 14*, 1677–1689. https://doi.org/10.5194/nhess-14-1677-2014

Brundson, C. H., Charlton, M., and Fotheringham, S. (1996). Geographically weighted regression: A method for exploring spatial nonstationarity. *Geographical Analysis 28*, 4

Charlton, M., and Fotheringham, S. (2010). *Geographically Weighted Regression. A Tutorial on using GWR in ArcGIS 9.3*. National Centre for Geocomputation, National University of Ireland Maynooth.

Charlton, M., and Fotheringham, S. (2009). *Geographically Weighted Regression.* White paper. National Centre for Geocomputation, National University of Ireland Maynooth.

Choi, J. K., Kim, K. D., Lee, S., and Won, J. S. (2010). Application of a fuzzy operator to susceptibility estimations of coal mine subsidence in Taebaek City, Korea. *Environmental Earth Sciences 59,* 1009–1022.

de Moura, E. N., and Procopiuck, M. (2020). GIS-based spatial analysis: Basic sanitation services in Parana State, Southern Brazil. *Environmental Monitoring and Assessment 192*(2), 1–13.

Fischer, M. M., Scholten, H. J., and Unwin, D. (2019). Geographic information systems, spatial data analysis and spatial modelling: An introduction. In: *Spatial Analytical Perspectives on GIS*. Routledge, pp. 3–20.

Fotheringham, A. S., Brunsdon, C., and Charlton, M. E. (2002). *Geographically Weighted Regression: The Analysis of Spatially Varying Relationships*. Chichester: Wiley.

Hurvich, C. M., and Tsai, C. L. (1993). A corrected Akaike information criterion for vector autoregressive model selection. *Journal of Time Series Analysis*, *14*(3), 271–279.

Jamal, A., Dhar, B. B., and Ratan, S. (1991). Acid mine drainage control in an opencast coal mine. *Mine Water and the Environment 10*(1), 1–16.

Khan, I., and Javed, A. (2012). Land use land cover change due to mining activities in Singrauli belt, Madhya Pradesh using remote sensing and GIS. *Journal of Environmental Research and Development 6*(3A), 834–843.

Knothe, S. (1984). *Forecasting the Influence of Mining (in Polish)*. Katowice: S´ląsk Publishing House.

Kratzsch, H. (1983). *Mining Subsidence Engineering*. Berlin: Springer.

Kuscu, S. (1991). Mine subsidence monitoring and importance of subject in Zonguldak coal region in Turkey. In: *Proceedings FIG International Symposium on Engineering Surveys*. Sofia, Bulgaria, pp. 225–236.

Lee, S., and Park, I. (2013). Application of decision tree model for the ground subsidence hazard mapping near abandoned underground coal mines. *Journal of Environmental Management 127*, 166–176.

Majumder, S., and Sarkar, K. (1994). Impact of mining and related activities on physical and cultural environment of singurali coalfield—A case study through application of remote sensing techniques. *International Journal of Remote Sensing 22*(1), 45–56.

Malinowska, A. (2014). Classification and regression tree theory application for assessment of building damage caused by surface deformation. *Natural Hazards 73*, 317–334.

Mathey, M. (2013). Addressing the challenges involved with abandoned underground coal mines in South Africa. In: *Proceedings of the 13th ISM Congress*. Aachen, Germany, pp. 113–123.

Oh, H. J., and Lee, S. (2011). Integration of ground subsidence hazard maps of abandoned coal mines in Samcheok, Korea. *International Journal of Coal Geology 86*(1), 58–72

Park, I., Lee, J., and Lee, S. (2014). Ensemble of ground subsidence hazard maps using Fuzzy Logic. *Central European Journal of Geosciences 6*(2), 207–218. https://doi.org/10.2478/s13533-012-0175-y

Popiołek, E. (2009). *Protection of Mining Terrains (in Polish)*. Cracow: AGH Publishing House.

Rodríguez-Espinosa, V. M., Aguilera-Benavente, F., and Gómez-Delgado, M. (2020). Green infrastructure design using GIS and spatial analysis: A proposal for the Henares Corridor (Madrid-Guadalajara, Spain). *Landscape Research*, *45*(1), 26–43.

Shang, R. K., Shiu, Y. S., and Ma, K. C. (2011). Using geographically weighted regression to explore the spatially varying relationship between land subsidence and groundwater level variations: A case study in the Choshuichi alluvial fan, Taiwan. *Proceedings IEEE International Conference on Spatial Data Mining and Geographical Knowledge Services*, IEEE, pp. 21–25

Sinclair Knight Merz Pty Ltd. (2014). *Subsidence from Coal Mining Activities*. Department of the Environment, Australian Government.

Stecchi, F., Mancini, F., Ceppi, C., and Gabbianelli, G. (2012). Vulnerability to ground deformation phenomena in the city of Tuzla (BiH): A GIS and multicriteria approach. In: *Nat Hazards*.

Tobler, W. (1970). A computer movie simulating urban growth in the Detroit region. *Economic Geography 46*(Supplement), 234–240.

Unwin, D. (2020). GIS and spatial statistical analysis. In: *Geographic Information Research*, CRC Press, pp. 399–411.

Verhagen, P. (2018). Spatial analysis in archaeology: Moving into new territories. In: *Digital Geoarchaeology*, Cham: Springer, pp. 11–25.

CHAPTER 7

A SPATIALLY EXPLICIT FRAMEWORK FOR COMPUTING THE QOL INDEX FOR PEOPLE IN THE CAUVERY RIVER BASIN

ABHISHEK GUPTA[1], RAMESH KRISHNAMURTHY[2], MOHAMMAD SAJID[1], and MOHAMMED ZAID QAMAR[1]

[1]*Interdisciplinary Department of Remote Sensing and GIS Application, AMU, Aligarh, Uttar Pradesh, India*

[2]*Wildlife Institute of India, Dehradun, Uttarakhand, India*

ABSTRACT

Quality of life (QoL) is described as a multidimensional idea incorporating social and economic well-being, inclusive of education, healthcare infrastructure, and accessibility to elementary facilities. QoL study over an area not only gives the idea of the human socioeconomic living condition but also gives us the sense of dependency of human life quality over other ecological and physical factors. This study maps the QoL of the administrative districts within the Cauvery River basin of southern India. It was found that Kerala scored highest as the district average of QoL of 0.8. District-wise, Kozhikode, Bangalore, and Coimbatore topped the QoL score (1, 0.99, and 0.95, respectively); Chamarajanagar, Chikkaballapura, and Ariyalur ranked lowest with scores 0, 0.18, 0.18, respectively. In addition, the high correlation of the sanitation subindex (0.78) and household latrine indicator (0.78) with the QoL index showed credence

Sustainable Development: A Geospatial and Statistical Perspective. Qazi Mazhar Ali, Rizwan Ahmad, Irfan Ali, & Prasenjit Chatterjee (Eds.)

DOI: 10.1201/9781779643247-7

over the relation and importance of the sanitation facility concerning QoL. Furthermore, the dimension of the living standard showed highly correlated and variance-holding subindexes (Sanitation (0.78); Household Cooking and Fuel (0.66); Occupancy (0.59 correlation), and −0.37, 0.36, and 0.38 variance scores, respectively). Dimension wise, Kerala scored highest in health (0.71) and environmental quality (0.76), whereas Tamil Nadu and Puducherry scored highest in education quality (0.65) and living standards (0.56), respectively.

7.1 INTRODUCTION

As humans start to evolve and populate, some issues start forming their place in their social and ecological framework, issues deriving from all the factors present in the whole ecosystem including natural phenomena like natural disasters and hazards, man-made phenomena like pollution or climate change, interhuman competition like communal conflicts, political conflicts. Furthermore, to compensate and adapt accordingly to these issues, humans are always subjected to creating methods, actions, and organizations for better livability and existence. The main factor or characteristic of a living being is its living quality. Living well-being is covered by a wide variety of factors including living environment, psychological state, physical state, socioeconomic conditions, and so on. In the current scenario as a developing country, problems such as segregation, neighborhood degradation, increased road traffic, socioeconomic deprivation, and inequities in health and well-being and healthcare accessibility have become a central issue for administrators and policymakers. These factors act as sure indicators for livability and the standard of living of the citizens. Also, to ensure better livability of citizens, new research is needed to quantify and assess these factors for better policymaking and provision of services. Quality of life (QoL) came out to be a multidimensional concept incorporating social and economic well-being, such as education, health, safety and security, access to basic amenities (Joshi et al., 2016), and living environment quality (Olivier et al., 2002). The studies concerning QoL are gaining interest in both developing and developed countries from a variety of disciplines such as planning, geography, sociology, economics, psychology, political science, behavioral medicine, marketing, and management (Andrew,

1999; Rao et al., 2012). It came out as a broad concept that assessed on various factors ranging from living conditions and employment to experience of life. It is usually represented by a multiple set of indicators such as income, deprivation rate, education attainment, employment rate, life expectancy, air quality, environmental quality. (Sapena et al., 2020). QoL research needs to investigate, identify, document, and report what defines the well-being of communities (Cobb, 2000). By mapping the variation of QOL, measured by carefully selected social indicators, we can pinpoint wherein a city or a state human suffering exists, obviating guesswork in the allocation of needed social assistance (Bederman and Hartshorn, 1984). Today, in the 21st century, we need to support and appreciate new research in this field, that is, geographic information system (GIS) application in social sciences to assist the high-paced technologically evolving species. The analysis of the society of human interactions and phenomena has now become easier using new tools and techniques, and one of them is the technique of GIS or geocomputation. As a part of this vast geography of the earth, we impart ourselves and our phenomenon in the form of geospatial data. The data gets processed in a GIS environment using statistical and informative science and gives an easily interpretable result for analysis and research with spatial reference. In addition, as India is a developing nation, the quality of life of the citizens highly depends on the infrastructure and its development in all aspects comprising their direct environment.

7.2 STUDY AREA

Cauvery is an easterly flowing river of peninsular India and flows through three states, i.e., Karnataka, Kerala, Tamil Nadu, and one union territory, Pondicherry. It is the fourth longest river in Southern India and the eighth longest in India with a length of around 800 km. It originates at an elevation of 1341 m at Talakaveri on the Brahmagiri range near Cherangala village of Kodagu district of Karnataka and meets or drains into the Bay of Bengal by forming the delta in eastern Tamil Nadu districts and Pondicherry. The basin lies between 75°27′ and 79°54′ east longitudes and 10°9′ and 13°30′ north latitudes. It has a total drainage area of 81,155 sq km of which 34,273 sq km lies in Karnataka, 43,856 sq km in Tamil Nadu, 2866 sq km in Kerala, and 160 sq km in the Union Territory of Pondicherry. Its important tributaries joining from the left are the Harangi, the Hemavati, the Shimshal, and the Arkavati, whereas the Lakshmana Tirtha, the Kabini, the Suvarnavati, the Bhavani, the Neil, and the Amaravati join from the right. Physio-graphically, the basin is divided into three divisions, that is, the Western Ghats, the plateau of Mysore, and the Delta; and bounded by the Western Ghats on the west, by the Eastern Ghats on the east and the south, and by the ridges separating it from the Krishna basin and Pennar basin on the north. The delta area is the most agricultural productive zone in the basin. A significant portion of the basin is enclosed with agricultural land accounting for 66.21% of the total area (Jena et al., 2020) (Figures 7.1 and 7.2).

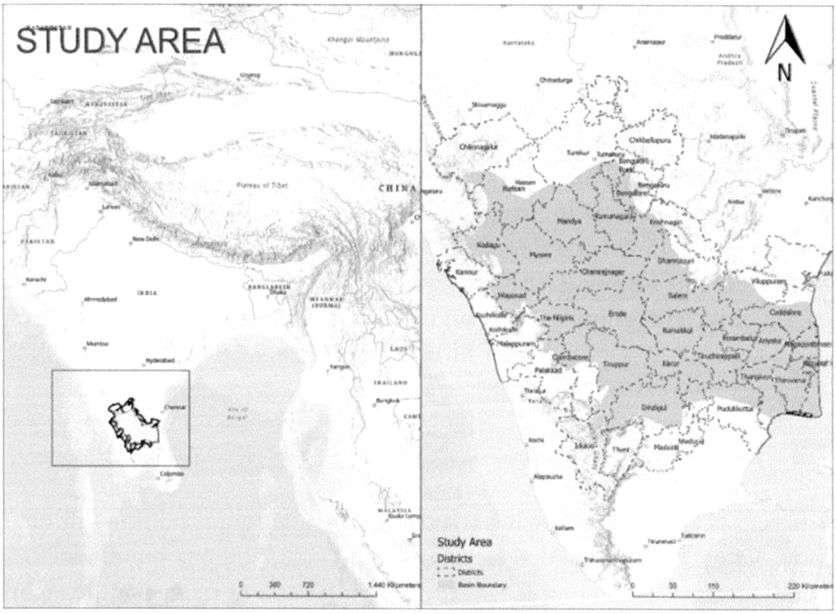

FIGURE 7.1 Study area delineated as Cauvery basin and districts.

The basin has a maximum elevation of about 2629 m. An elevation of about 2000–3000 m is observed in 1% of the total geographical area of the basin. Around 32% of the basin area falls under the 750–1000 m elevation zone (Cauvery Basin V2.0 IWRIS (India Water Resource Information System), 2014) (Figure 7.3, Table 7.1).

FIGURE 7.2 Cauvery basin Index map (as per Survey of India (SOI)).

7.3 DATASETS

7.3.1 HOUSE LISTING AND HOUSING CENSUS DATA-2011

This dataset was used to extract the basic statistics of households and houses (total number of houses and household), condition and occupancy of houses, availability of essential resources (drinking water sources, main lighting source, kitchen condition, and the fuel used in cooking), and availability of personal and community assets (vehicle, mobile/laptop, shops, hotel, workshops) for the QoL analysis.

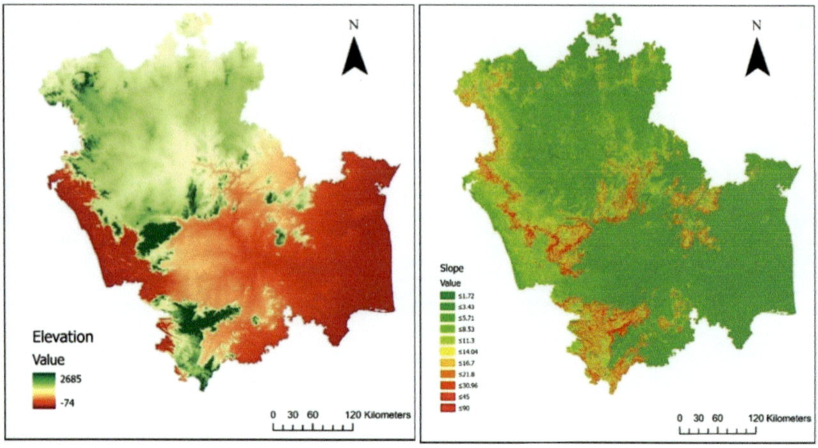

FIGURE 7.3 Mean elevation and slope steepness.

TABLE 7.1 States and Districts in the Basin

Tamil Nadu		Karnataka	Kerala	Puducherry
Ariyalur	Tiruchirappalli	Chikmagalur	Kannur	Karaikal
Coimbatore	Perambalur	Tumkur	Wayanad	
Cuddalore	Nagapattinam	Bangalore Rural	Kozhikode	
Dharmapuri	Thiruvarur	Bangalore	Malappuram	
Dindigul	Thanjavur	Mandya	Palakkad	
Erode	Pudukkottai	Hassan	Idukki	
The Nilgiris	Theni	Kodagu		
Salem	Tiruppur	Mysore		
Villupprum	Madurai	Chmarajnagar		
Namakkal	Krishnagiri	Ramanagara		
Karur		Chikkaballapura		

7.3.2 OPENSTREETMAP DATA

OpenStreetMap is built by a community of mappers that contribute and maintain data present at the street level. The road network data was downloaded and verified with the SOI Toposheets. *The data is freely available for download at OpenStreetMap*

7.3.3 NATIONAL HEALTH MISSION REPORT 2018–2019

The data was extracted from the latest report published under the National Health Mission. The data was used to formulate the variable for the health-care infrastructure.

7.3.4 DISTRICT REPORT CARD REPORT 2016–2017

In this chapter, elementary education (primary and upper primary level) was taken as an exclusive indicator; also, the higher education data was not available in a collectible format at the district level. The data was collected from the district report card report 2016–2017.

7.3.5 IWRIS

IWRIS Groundwater portal (https://indiawris.gov.in/wris/) was used to download the net groundwater availability data. The data is available in tabular at different scales from country to taluk level.

7.4 METHODOLOGY

7.4.1 SELECTION OF INDICATORS

At the district administrative level, the framework was created based on prior research studies and available datasets at the appropriate level. As per the developing scenarios in India, the variables were broadly categorized into four dimensions defining QoL, that is, Health Quality, Education Quality, Living Standards, and Environment Quality (Table 7.2).

7.4.2 DEVELOPMENT OF THE INDICATORS/NORMALIZATION OF RAW INDICATOR VALUES

A total of 66 variables were taken to compute the cumulative QoL index. The variables had different units, so normalization was carried out using the standard formula:

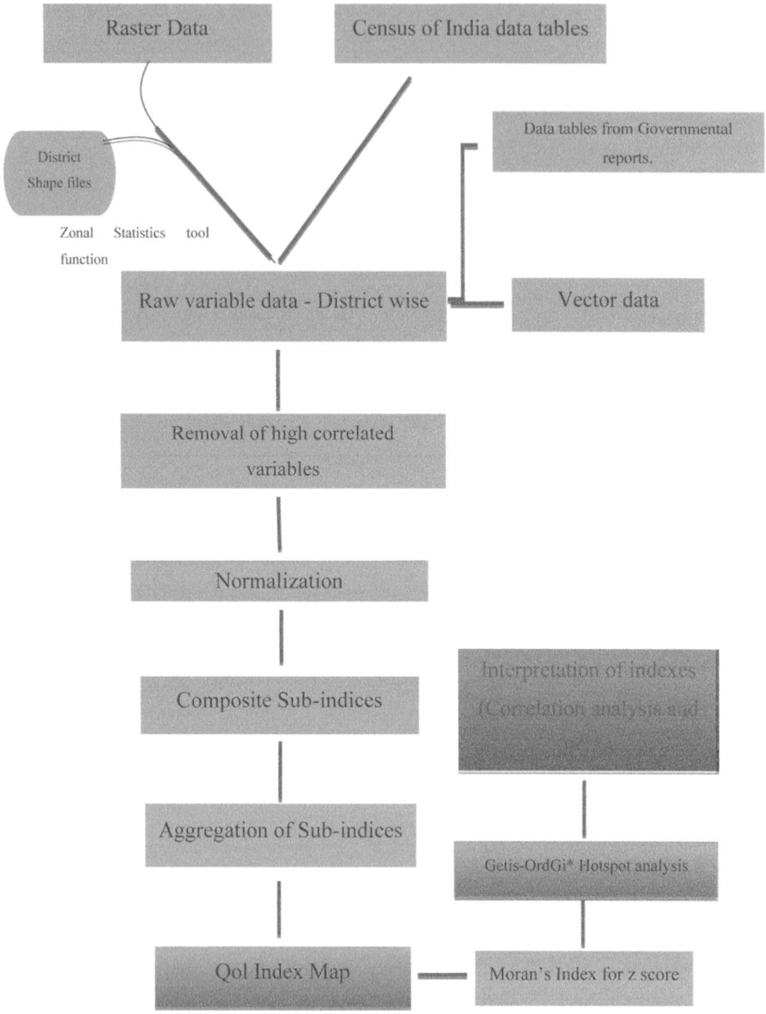

FIGURE 7.4 Correlation of QoL index with indicators.

TABLE 7.2 Subindexes and Indicators

Sr. No	Subindexes/Indicators	Description/Criteria
	Living Standards	
I	**Household Lighting**	
1	HL-Electricity	Percentage of Total Households
2	HL-Solar energy	Percentage of Total Households

TABLE 7.2 *(Continued)*

Sr. No	Subindexes/Indicators	Description/Criteria
3	HL-Other Oil	Percentage of Total Households
4	HL-No Lighting	Percentage of Total Households
II	**Household Cooking Fuel**	
5	Household kitchen	Percentage of Total Households
6	HF-Crop residue	Percentage of Total Households
7	HF-Cowdung	Percentage of Total Households
8	HF-Coal, Lignite, Charcoal	Percentage of Total Households
9	HF-LPG/PNG	Percentage of Total Households
10	HF-Electricity	Percentage of Total Households
11	HF-Biogas	Percentage of Total Households
12	No cooking	Percentage of Total Households
III	**Drinking Water Accessibility**	
13	DW-Within premise	Percentage of Total Households
14	DW-Treated tap water	Percentage of Total Households
15	DW-Improved source	Percentage of Total Households
IV	**Sanitation**	
16	HS-Latrine within premises	Percentage of Total Households
17	HS-Batinng within premises	Percentage of Total Households
18	HS-Closed Drainage	Percentage of Total Households
V	**Houses Occupancy**	
19	HH-Residential	Percentage of Total Census Houses*
20	HH-Occupied Locked	Percentage of Total Census Houses
VI	**Houses Condition**	
21	Good	Percentage of Total Census Houses
22	Dilapidated	Percentage of Total Census Houses
VII	**Household Assets**	Percentage of Total Census Houses
23	Banking services	Percentage of Total Households*
24	Radio/Transistor	Percentage of Total Households
25	Household with TV, Computer/Laptop, Telephone/Mobile phone and Scooter/Car	Percentage of Total Households
26	None of the assets (comp/laptop, telephone, bicycle/scooter/car)	Percentage of Total Households
VIII	**Demographic and Social Stability**	
27	Population-Density	Population per sq. km

TABLE 7.2 *(Continued)*

Sr. No	Subindexes/Indicators	Description/Criteria
28	Sex Ratio	Number of female per male
29	Dependent population	Total dependent population over working population
30	Crime Cases	Total crime cases (composite of IPC and SLL cases)
IX	**Community Assets**	
31	Shop/Office	Shop/office per sq. km
32	Road Network Density	Total Area of roads per sq. km
33	Heritage and Cultural Sites	Number of sites
	Health	
X	**Health Infrastructure**	
34	Government Health Centers	Health centers over every 100 persons
35	Vaccinated population COVID	Percentage of total Population
36	Death Ratio COVID	Death by total cases of COVID
37	Bed Allocation COVID	Total beds per person
XI	**Family Planning**[+]	
38	Using FP Methods	Percentage of Household/Family
39	Total Unmet needs	Percentage of Household/Family
XII	**Women and Child Care**[*]	
40	Deliveries in Health facility	Percentage of total deliveries
41	Antenatal care from doctors	Percentage of total deliveries
42	Financial assistance under JSY	Percentage of total deliveries
43	Children receiving all vaccination	Percentage of total child population
44	Children having anemia	Percentage of total child population
45	Women having anemia	Percentage of total women population
46	Women with low BMI	Percentage of total women population
47	Women with high BMI	Percentage of total women population
	Environment	
XIII	**Geographic Suitability**	
48	Mean Slope	Mean slope under the administrative boundary (district)
49	Mean Elevation	Mean Elevation under the administrative boundary (district)

TABLE 7.2 *(Continued)*

Sr. No	Subindexes/Indicators	Description/Criteria
50	Urban Heat Island	Percentage of area of UHI of the total area
XIV	**Ecological Resources**	
51	Mineral Sites	Number of mineral sites
52	Forest Cover	Total Forest cover in percentage of total area
53	Increase in forest cover	Total change in sq.km
54	Ground Availability	Net Ground water availability
55	Surface Water Cover	Percentage of water cover of the total area
	Education	
XV	**Infrastructure**	
56	Schools approachable by all weather roads	Percentage of total schools
57	Schools with boundary wall	Percentage of total schools
58	Schools with playground	Percentage of total schools
59	Schools with girls toilet	Percentage of total schools
60	Schools provided MDM	Percentage of total schools
61	Schools with electricity	Percentage of total schools
62	Schools with computers	Percentage of total schools
63	Schools with kitchen shed	Percentage of total schools
XVI	**Other indicators**	
64	Overall Literacy rate	Absolute
65	Schools per sq km	Schools per sq. km
66	Pupil teacher ratio	Number of students per teacher

*A "Census house" is a building or part of a building used or recognized as a separate unit because it has a separate main entrance from the road or common courtyard or staircase, etc.

*A "household" is usually a group of persons who normally live together and take their meals from a common kitchen.

+Percentage of last in the last five years for variables relating to deliveries (Antenatal care, financial assistance)

+Percentage of currently married women aged 15–49 years using any contraceptive method, that is, any modern method, female sterilization, male sterilization, any modern spacing method, and any traditional method, according to the district.

$$I = \frac{I_a - I_{min}}{I_{max} - I_{min}}$$

Here, I is the normalized value of the class, I_a is the actual value of the same variable, and I_{min} and I_{max} represent the minimum and maximum values, respectively, of the variable.

The normalized variables were then checked for correlation using Spearman correlation in *R* studio to check the highly correlated variables. The redundant variable was then removed from the further analysis.

The variables were then averaged to form the respective logical composites for subindexes using the standard averaging formula:

$$A = \frac{1}{n}\sum_{i=1}^{n} a_i$$

Here, A is the average, n is the number of variables used to create composites, and a_i is the value of the variable.

7.4.3 CONVERSION OF COMPOSITE VALUES TO SUBINDEXES

The minimum and maximum allowable values—also known as lower and upper bounds—must be specified before constructing indexes whose values can vary between the chosen scale, such as 0–10 or 0–100 (here 0–1).

The basic formula for converting an indicator value (V) into an index score (I) is

$$i = \frac{V - min\_value}{nax\_value - min\_value}\text{'}$$

where *min_value* is the minimum admissible value (lower bound) and *max_value* is the maximum admissible value (upper bound).

In a few cases, indicators and criteria point in opposite directions. For example, a high crime signifies a low (rather than high) level of social stability. In these cases, the following formula was used:

$$i = \frac{max\_value - V}{max\_value - min\_value}.$$

The nonweighted calculated composite for the QoL was formed by averaging the normalized values of subindexes to produce an average total of the indexes.

7.4.4 CALCULATING QOL INDEX

QoL Index was calculated by the above-stated standard min–max normalization of the composite of the 16 subindexes discussed under four dimensions: Living Standards (Household Lighting, Household Cooking and Fuel, Drinking Water Accessibility, Sanitation, Household Occupancy, House Condition, Household assets (HS), Demographic and social stability, and community assets); Health Quality (Health infrastructure, Family Planning, and Women and childcare); Environment Quality (Geographic Suitability and Ecological resources); and Education Quality (Education Infrastructure and other education indicators).

7.4.5 STATISTICAL ANALYSIS

The Pearson correlation was used to conduct a cross-correlation study between the QoL index and the subindexes to determine their reliance and connection. Principal component analysis (PCA) was also used to investigate the variance. The statistical analysis was carried out in *R*, while the hotspot analysis was carried out in ArcGIS Pro.

7.5 RESULTS AND DISCUSSION

The QoL index revealed some clustering of high QoL scores visually around the western part of the research area, physically in the Western Ghats and central plateau covering Kerala and Tamil Nadu districts. Bangalore, Karnataka's urban agglomerate, received the highest QoL after Kozhikode, indicating the existence of higher and better household cooking fuel, sanitation, and house occupancy than other districts in the state (Figure 7.4, Table 7.3).

7.5.1 CORRELATION ANALYSIS WITH INDICATORS

The computed QoL index, when run with Pearson correlation with the raw indicators, showed a high correlation with household latrine within premises, drinking water availability within premises, women with less body mass Index or underweight women, and overall literacy rate, with correlation coefficients of 0.78, 0.73, −0.71, and 0.68, respectively. A negative connection indicates that women's health and nutrition have a detrimental impact on QoL.

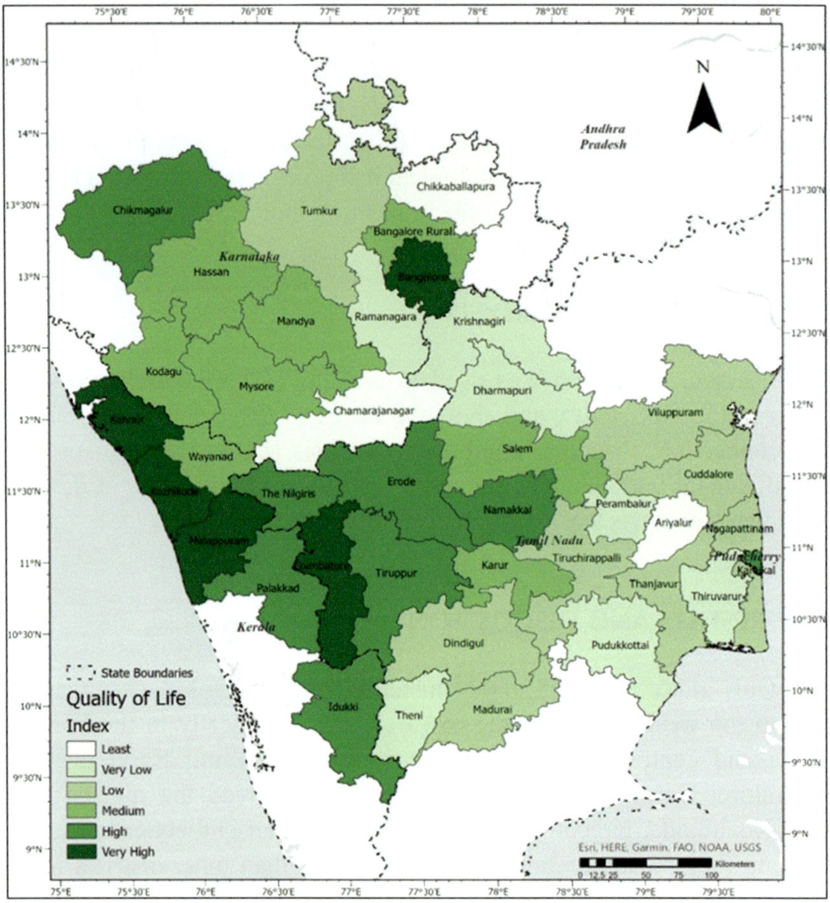

FIGURE 7.5 Quality of life index map.

TABLE 7.3 Quality of Life Index (Sorted Descending)

Level	State/UT	Name	Qol Index
DISTRICT	Kerala	Kozhikode	1
DISTRICT	Karnataka	Bangalore	0.99
DISTRICT	Tamil Nadu	Coimbatore	0.95

TABLE 7.3 *(Continued)*

Level	State/UT	Name	Qol Index
DISTRICT	Kerala	Kannur	0.9
DISTRICT	Kerala	Malappuram	0.88
DISTRICT	Tamil Nadu	Erode	0.77
DISTRICT	Kerala	Palakkad	0.76
DISTRICT	Puducherry	Karaikal	0.73
DISTRICT	Tamil Nadu	Tiruppur	0.71
DISTRICT	Tamil Nadu	The Nilgiris	0.69
DISTRICT	Kerala	Idukki	0.66
DISTRICT	Karnataka	Chikmagalur	0.66
DISTRICT	Tamil Nadu	Namakkal	0.65
DISTRICT	Karnataka	Mysore	0.63
DISTRICT	Kerala	Wayanad	0.61
DISTRICT	Karnataka	Mandya	0.59
DISTRICT	Tamil Nadu	Karur	0.58
DISTRICT	Karnataka	Kodagu	0.56
DISTRICT	Karnataka	Bangalore Rural	0.56
DISTRICT	Karnataka	Hassan	0.53
DISTRICT	Tamil Nadu	Salem	0.51
DISTRICT	Tamil Nadu	Nagapattinam	0.49
DISTRICT	Tamil Nadu	Tiruchirappalli	0.46
DISTRICT	Tamil Nadu	Dindigul	0.43
DISTRICT	Tamil Nadu	Thanjavur	0.43
DISTRICT	Tamil Nadu	Cuddalore	0.41
DISTRICT	Tamil Nadu	Viluppuram	0.39
DISTRICT	Tamil Nadu	Madurai	0.37
DISTRICT	Tamil Nadu	Pudukkottai	0.35
DISTRICT	Tamil Nadu	Thiruvarur	0.33
DISTRICT	Karnataka	Ramanagara	0.32
DISTRICT	Tamil Nadu	Perambalur	0.32
DISTRICT	Tamil Nadu	Theni	0.29
DISTRICT	Tamil Nadu	Dharmapuri	0.28
DISTRICT	Tamil Nadu	Krishnagiri	0.28
DISTRICT	Tamil Nadu	Ariyalur	0.18
DISTRICT	Karnataka	Chikkaballapura	0.18
DISTRICT	Karnataka	Chamarajanagar	0

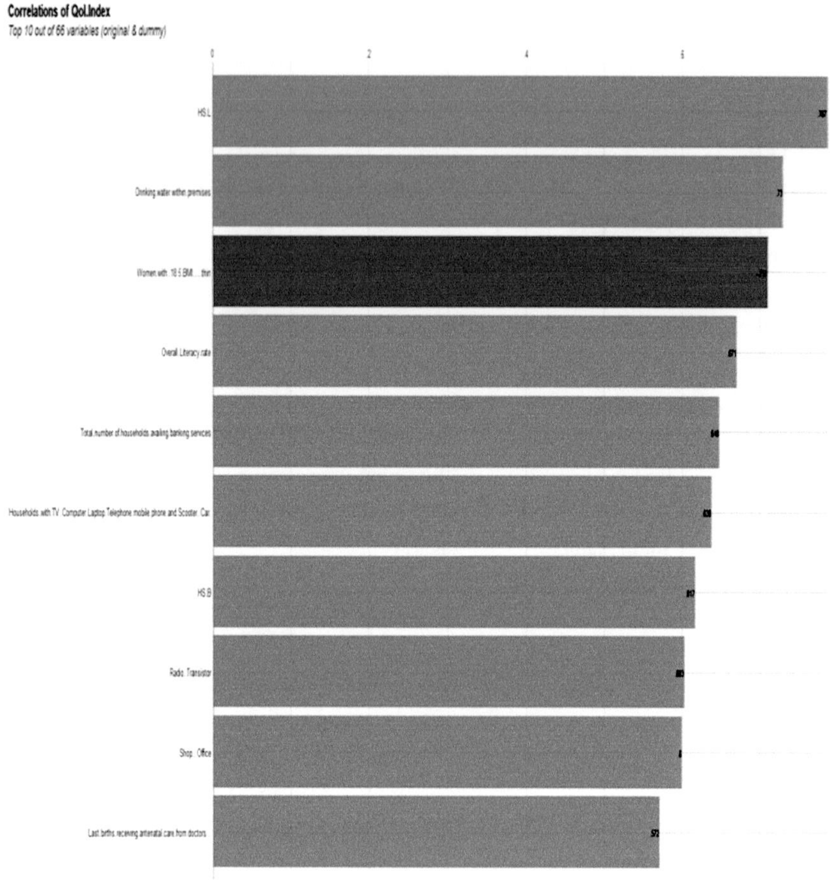

FIGURE 7.6 Correlation of QoL Index with indicators.

7.5.2 CORRELATION ANALYSIS WITH SUBINDEXES

The computed QoL index on running with Pearson correlation with other subindexes gives the following results (Figure 7.5, Table 7.4).

This clearly shows a high positive correlation of HS and sanitation with the overall QoL index.

HS include variables: Banking services, Radio/Transistor, Households with TV, Computer/Laptop, Telephone/mobile phone, and Scooter/Car, None of the assets (comp/laptop, telephone, bicycle/scooter/car).

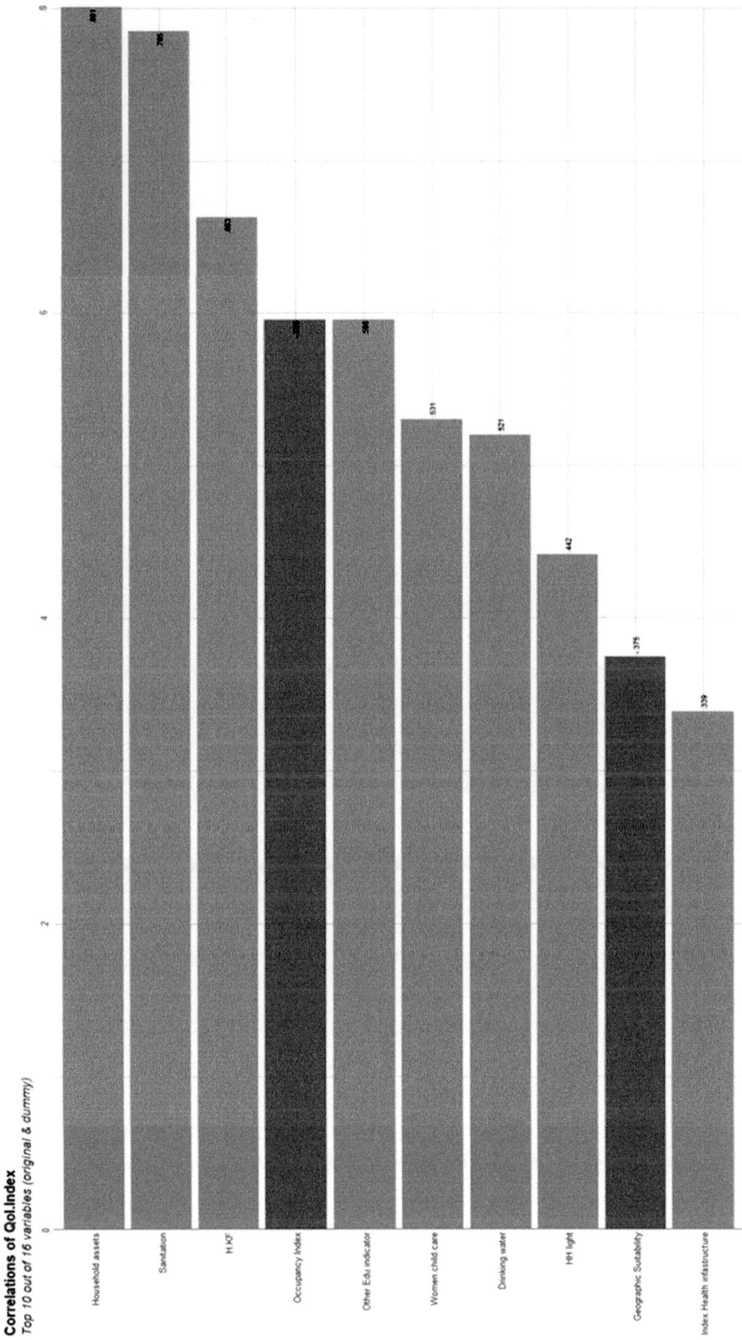

FIGURE 7.7 Correlation of QoL with subindexes.

TABLE 7.4 Top Five Correlations with Subindexes

Correlation Coefficient	Subindex
Household Assets	0.801
Sanitation	0.785
Household Cooking and Fuel	0.663
Occupancy	−0.596
Other Education indicators	0.596

Sanitation includes HS—latrine within premises, bathing within premises, and closed drainage.

7.6 CONCLUSION

The study's main goal was to create a framework for assessing QoL, and the conceptual framework was based on the district's social and developmental situation. The QoL Index was created under four broad dimensions (Health Quality, Living Standards, Environment Quality, and Education Quality) with 16 subindexes that are significant for human growth and QoL. The key finding of the study was that the dimension of living standard showed highly correlated and variance-holding indexes and indicators such as Sanitation (0.78); Household Cooking and Fuel (0.66); Occupancy (0.59 correlation), and −0.37, 0.36, and 0.38 variance scores, respectively. The finding and the methodology can be further used for other integrated research at the district level like the correlation of different fronts with the QoL. This study, (QoL Assessment) examines and signifies the impact of various assets, services, and social circumstances on people's QoL, as well as provides frameworks for the optimal distribution of services to people on a local level. Also, in today's scenario, as river basins are subjected to urbanization, resulting in indirect environmental exploitation, this study will aid policymakers and researchers in achieving sustainable development goals with an eye to environmental conservation, as well as the mainframe structure of policies operating on the ground.

KEYWORDS

- **indicators**
- **IWRIS**
- **OpenStreetMap**
- **principal component analysis**
- **quality of life**
- **river basin**
- **subindexes**

REFERENCES

Andrew, P. J., and Mayer, B. (1999). Enzymatic function of nitric oxide synthases. *Cardiovascular Research, 43*(3), 521–531.

Bederman, S., and Hartshorn, T. (1984). Quality of life in Georgia: The 1980 experience. *Southeastern Geographer 24*(2), 78–98.

Blum, H. (1981). *Planning for Health*. New York: Human Sciences Press.

Censusindia.gov.in. (2021). *Census of India Website: Office of the Registrar General & Census Commissioner, India*. Retrieved from https://censusindia.gov.in/2011-common/censusdata2011.html

Cobb, C. (2000). Measurement tools and the quality of life. *Redefining Progress*. Retrieved from https://www.academia.edu/2130342/Measurement_tools_and_the_quality_of_life.

Dissanayake, D., Morimoto, T., Murayama, Y., Ranagalage, M., and Perera, E. (2020). Analysis of life quality in a tropical mountain city using a multi-criteria geospatial technique: A case study of Kandy City, Sri Lanka. *Sustainability 12*(7), 2918.

Gholami, S., and Srikantaswamy, S. (2009). Statistical multivariate analysis in the assessment of river water quality in the vicinity of KRS Dam, Karnataka, India. *Natural Resources Research 18*(3), 235–247.

Ibis.geog.ubc.ca. (2021). *Spatial Analysis*. Retrieved from https://ibis.geog.ubc.ca/courses/geob370/students/class10/bcmckay/www/Site/Spatial_Analysis.html

Jena, P., Kasiviswanathan, K., and Azad, S. (2020). Spatiotemporal characteristics of extreme droughts and their association with sea surface temperature over the Cauvery River basin, India. *Natural Hazards 104*(3), 2239–2259.

Joshi, R., and Ahmed, S. (2016). Status and challenges of municipal solid waste management in India: A review. *Cogent Environmental Science, 2*(1), 1139434.

Lawrence, R. (2002). Our fragile world: Challenges and opportunities for sustainable development. *International Journal of Sustainability in Higher Education*, *3*(2), 182.

Li, G., and Weng, Q. (2007). Measuring the quality of life in city of Indianapolis by integration of remote sensing and census data. *International Journal of Remote Sensing* *28*(2), 249–267.

Lo, C. (1997). Integration of landsat thematic mapper and census data for quality of life assessment. *Remote Sensing of Environment* *62*(2), 143–157.

Lotfi, S., and Solaimani, K. (2009). An assessment of urban quality of life by using analytic hierarchy process approach (case study: Comparative study of quality of life in the North of Iran). *Journal of Social Sciences* *5*(2), 123–133.

Olivier, J. G. J. L. J., Brandes, Peters, J. A. H. W., Coenen, P. W. H. G., van der Hoek, K. W., Montfoort, J. A., Peek, C. J., Nijdam, D. S., Vreuls, H. H. J., and Zonneveld, E. A. RIVM Report 773201 006/2002 Greenhouse Gas Emissions in the Netherlands 1990 2000. National Inventory Report 2002.

Rao, R. V., Descamps, O., John, V., and Bredesen, D. E. (2012). Ayurvedic medicinal plants for Alzheimer's disease: A review. *Alzheimer's Research & Therapy*, *4*, 1–9.

Toro, S. D., Isabel, M., Calabuig, J. M., Ferrer Sapena, A., and Peset Mancebo, M. F. (2020). Trabajar con datos abiertos en tiempos de pandemia: uso de covidDATA-19. *El Profesional de la Información*, *29*(4), 1–13.

CHAPTER 8

A REMOTE SENSING AND GIS-BASED TWO DECADAL SPATIOTEMPORAL ANALYSIS OF DELHI REGION DUE TO VARIABILITY IN CLIMATE

WAIZA KHALID[1], SAMREEN FATIMA[1], ANAND SHARMA[2], and MOHAMMAD SAJID[1]

[1]*Interdisciplinary Department of Remote Sensing and GIS Application, Aligarh Muslim University, Aligarh, Uttar Pradesh, India*

[2]*Regional Meteorological Centre, Indian Meteorological Department (IMD), Lodhi Road, New Delhi, India*

ABSTRACT

Climate is the long-term pattern of conditions in the atmosphere, their changes, and extremes in a region. The combined effect of all of these elements is represented in the variance of climate throughout time scales. Climate change and fluctuation have a bearing on land use/land cover (LULC) across the region. This study looks at how variations in climate and LULC have changed throughout time. LULC change is possibly the most visible form of worldwide changes in the environment at both time and space scales, with significant implications for our daily lives. Anthropogenic and environmental forces both have an important influence on the course of changes in LULC. In this study, for observing climate change in the last two decades 1999–2019, meteorological data is used with five climatic parameters, that is, temperature, rainfall, wind velocity,

Sustainable Development: A Geospatial and Statistical Perspective. Qazi Mazhar Ali, Rizwan Ahmad, Irfan Ali, & Prasenjit Chatterjee (Eds.)

DOI: 10.1201/9781779643247-8

sunshine hours, and relative humidity. The average was taken per year and variability is shown in the graphic form. Then LULC of Delhi is analyzed for the time period of the last two decades (1999–2019). Landsat 5 TM for 1999 and 2009, and Landsat 8 OLI and TIRS for 2019 were employed in this investigation, all with a spatial resolution of 30 m. For preparing LULC, supervised classification using the maximum likelihood classifier is used. In ArcGIS software, a change detection algorithm is used to detect shifts in LULC during the past two decades, from 1999 to 2019. There are four main groups defined, but the two primary classes are vegetation and built-up area. During the past 20 years, there has been an enormous shift. Vegetation cover has increased from 40.99% to 41.1%, built-up area has rise from 25.37% to 42.25%, water bodies have declined from 3.59% to 1.9%, and open scrub or bare land or the land on which work process is under process has declined from 30.03% to 14.06%. The classification accuracy has been established from Kappa statistics showing 93.41% for the year 1999, 96.66% for the year 2009, and 82.08% for the year 2019. It is clear from this study that the climate has a direct link with LULC, as the climate changes so does the land.

8.1 INTRODUCTION

Climate is the overall pattern of long-term fluctuations and extremes in atmospheric conditions in a certain place. The combined effect of all of these elements is represented in the variance of climate throughout time scales (Budhiraja et al., 2017). Changes in the climate and fluctuation have an impact on land use/land cover (LULC) across the region. While the climate generally changes slowly, we do experience short-term swings on seasonal or multiseasonal temporal spans. There are numerous factors that can cause temperature to fluctuate around an average without causing the historical mean to alter. Climate variability is a phenomenon that occurs throughout time intervals ranging from months to 30 years. When we talk about climatic variability, we are mostly talking about natural phenomena that affect the atmosphere, such as precipitation humidity levels, temperatures, light hours, pressure in the atmosphere, the direction of the wind, and the speed of the wind, among other things. According to the World Meteorological Organization (WMO), climatic variability is defined as differences in the average condition and additional climate statistics across all time and space scales, extending beyond isolated meteorological incidents. In simpler terms, variability is the range of climate compared to its average. Climate variability encompasses a range of changes that can affect temperature, precipitation patterns, and other factors across timeframes ranging from a few weeks to several decades. The relationship between climate change and land use is intricate. First, the global concentration of greenhouse gases is influenced by land cover, which is determined by land use practices. Second, although a major contributor to climate change, changes in LULC can also result from a changing climate. A century or so ago, the concepts of climate variability and anthropogenic climate change resulting from activities such as deforestation and reforestation preoccupied geographers, meteorologists, and climatologists (https://doi.org/10.1146/annurev.energy.28.050302.105459). Remote sensing and geographic information systems have been used as the foundation for developing affordable technologies for the wide-scale assessment of a variety of environmental monitoring issues, such as forestry management, ecological restoration, and land use and condition assessment (https://www.britannica.com/place/Delhi). The physical surface temperature (ST) is one of the most critical environmental elements influencing the passage of both energy and matter on the surface of the planet and the atmosphere below it. Information regarding possible climate change is anticipated to be obtained from a continuous measurement of this metric. The only method available for determining ST at the regional scale is satellite data. Research highlights solar variability as a catalyst or influencer of global warming. The climate, which includes measurements and modeling of temperature, precipitation, and other variables, is one of the solar forcing variables. There is proof of climate change based on the examination of historical land ST.

8.2 STUDY AREA

8.2.1 GEOGRAPHICAL SET UP

Delhi is bounded on the east by Uttar Pradesh, on the southwest borders by Rajasthan, and on its northwest and south by Haryana. It is located in northern India at 23.38°N and 77.13°E. The three main parts of the metropolis city are the Yamuna stream plain, the Delhi Ridge, and the plains of the Ganges. Although there is rich and fertile alluvial soil on the flood plain, which makes agriculture viable there, the flood line can occasionally rise to a staggering 318 m, which would destroy nearby crops. Aravalli Hill range forms the dominant ridge that encircles the city from the south to the west, northwest, and northeast. The sacred Yamuna River runs through Delhi, and the Hindon River separates Ghaziabad from the city's eastern outskirts (https://www.epa.gov/scram/meteorological-data-and-processors). The Yamuna River is crossed by numerous road bridges and Metro Subways. Delhi has a total area of 1513 kms^2, of which 783 are in the rural and 730 are in the urban sectors. The area is roughly 48.48 km wide and 51.9 km long. The area is prone to earthquakes because it is located in seismic zone IV.

8.3 CLIMATE OF DELHI

Delhi's climate is characterized by excessive dryness and brutally hot summers. Except during the monsoon season, when an easterly to the south-easterly entry of oceanic air provides rain and increases moisture, it is linked with a general presence of regional air that sweeps in from the western or northwestern. The summertime lasts from the middle of March until the final day of June, with the highest temperatures frequently reaching around 37°C and lows in the high 25°C. It is distinguished by numerous thunderstorms and storms, particularly in April and May. The season of monsoons typically starts in July and lasts till the close of September. During these months, Delhi obtains the majority of its rainfall—around 23–600 mm. October as well as November are transitional months from rainy to wintertime conditions. From the end of November to the middle of March, the arid winter season lasts. January is the coldest month, with temperatures reaching the low 21°C and lows in the mid 7°C. Here, the weather is referred to as "continental climate." The combination of the bitter cold of the Himalayas and the intense aridity of Rajasthan's deserts is what some people consider to be Delhi's terrible climate. Here, temperatures often reach 40°C from April to October. In December and January, the temperature gets quite close to zero. Over these two months, heavy fog shrouds Delhi. February–April and September–November are the most pleasant months. The range of temperatures is 20°C–30°C. August and July are the wet seasons. Over 67 cm, or 27 inches, of rain falls on Delhi annually on average. The Monsoon typically arrives in Delhi on June 29 of each year. Summertime water shortages are common because of the hot weather. During this time, there are also spells of rain and dust, with a heat wave peaking in May (Figure 8.1).

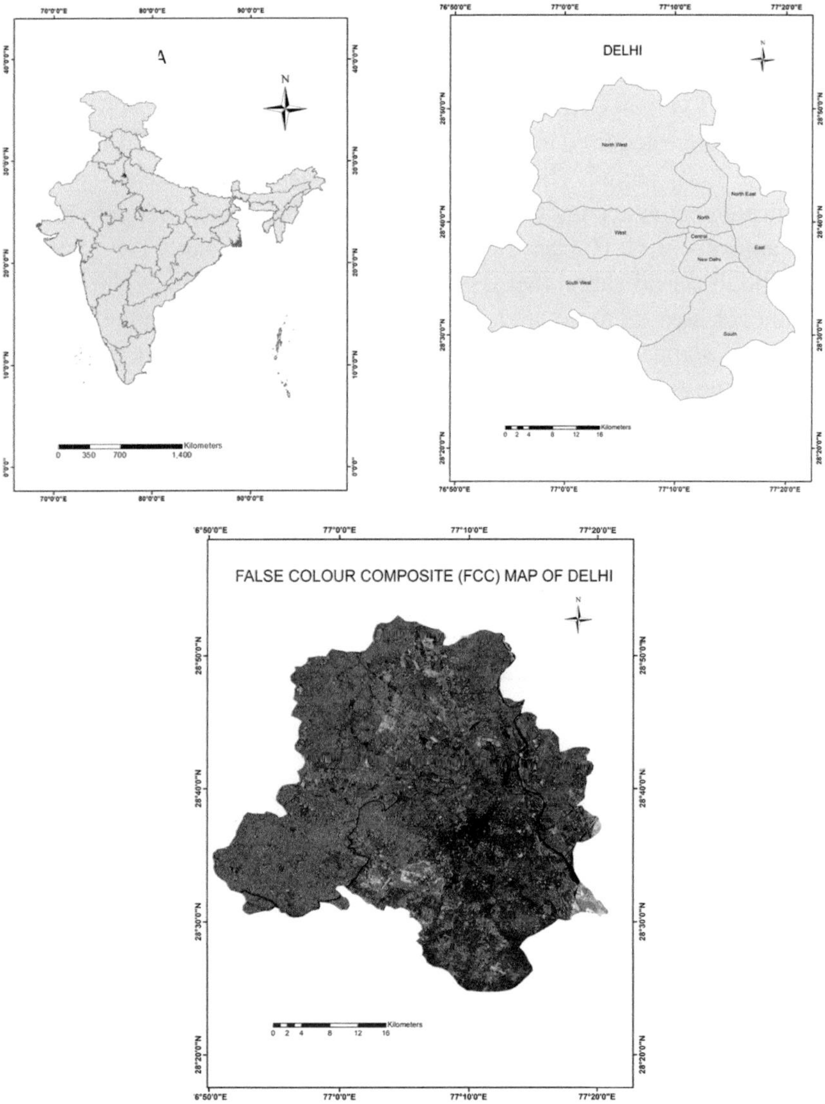

FIGURE 8.1 Study area map.

8.4 METHODOLOGY

The methodology employed in this study involved the collection and analysis of both meteorological and satellite data. Meteorological data comprising climatic parameters such as temperature, rainfall, wind velocity, sunshine hours, and relative humidity were gathered for the period 1999–2019. These parameters were analyzed to examine trends and variability, with variability graphs created to illustrate the temporal changes. Satellite data, including Landsat 5 TM images for 1999 and 2009 and Landsat 8 OLI & TIRS images for 2019, were acquired to analyze land use and land cover (LULC) changes over the study period. The satellite images underwent geometric correction to ensure spatial accuracy and were mosaicked to create seamless datasets for analysis. LULC classification was performed using the maximum likelihood classification (MLC) algorithm, which categorized the land into distinct classes for the years 1999, 2009, and 2019. To ensure the reliability of the classification, accuracy assessments were conducted for all three LULC maps (Parveen and Sarkar, 2018; Roy, 2011).

The climatic data were further analyzed to examine variability trends over the two decades. The findings from the climatic variability analysis were integrated with the LULC data to evaluate the impact of climatic changes on LULC transformations. This comprehensive analysis provided insights into the interplay between climatic factors and land use changes in the study area over the specified timeframe (Figure 8.2).

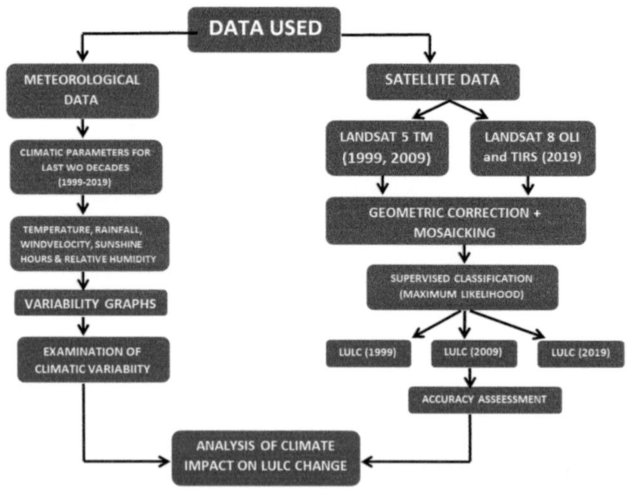

FIGURE 8.2 Methodology flowchart.

8.5 RESULT AND DISCUSSION

8.5.1 YEAR 1999

The Delhi surface cover map from 1999 shows the westward growth of urbanized accumulated regions (Figure 8.3) (Balogun et al., 2011; Chen et al., 2005; Hageback et al., 2005; Hassan et al., 2016). The northern region experiencing the most growth and the southern region experiencing relatively less growth. The Yamuna River may be seen flowing through the water bodies and little blue patches represent water. There are a growing number of green spaces in Delhi's south-eastern region, representing forests, vegetation, crops, and other natural areas, whereas the city's north, west, and southern-west sections have less greenbelts. Some areas in Delhi's north-western, western, and major south-western regions were under cultivable land area, along with the presence of wasteland (Tables 8.1 and 8.2).

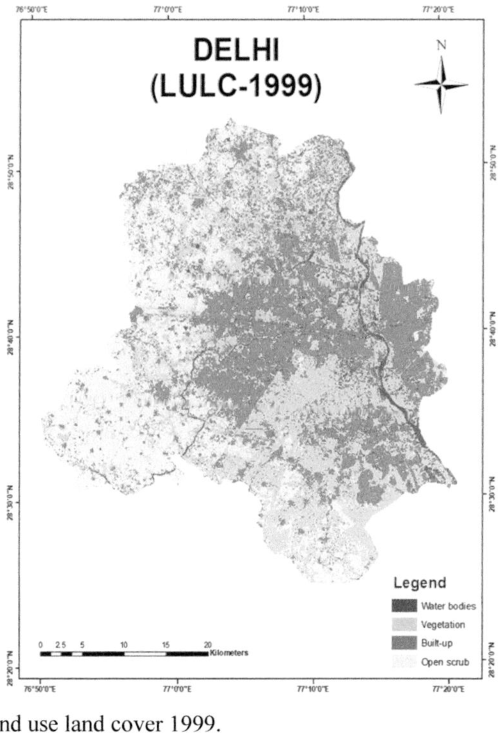

FIGURE 8.3 Land use land cover 1999.

Classification accuracy = 94.68%, Kappa statistics = 0.934%, Kappa statistics percentage = 93.41%

TABLE 8.1 Area for LULC 1999

Classes	Area (sq km)
Water bodies	54.33
Vegetation	620.31
Built-up	383.86
Open scrub	454.42

TABLE 8.2 Confusion Matrix for the Year 1999

	ID	Water Bodies	Vegetation	Built-Up	Open Scrub	Total
Water bodies	1	155	5	2	0	162
Vegetation	2	5	155	11	6	177
Built-up	3	0	0	142	0	142
Open scrub	4	0	0	5	154	159
Total		160	160	160	160	640

8.5.2 YEAR 2000–2009

The map of 2009 of Delhi shows that the settlement area has expanded enormously over north and northwest and still expanding toward north and south due to the rapid pace of urbanization (Figures 8.4 and 8.5). The vegetation area has been almost cleared up on the north side of Delhi due to the rise in settlement area on that side and some of the water bodies have dried due to climatological and anthropogenic factors (Tables 8.3 and 8.4). Some patches of vegetation on the south side of Delhi are converted into open scrub for future projects. The built-up region has expanded in every direction, but it is more obvious to the western and northern sides of the metropolitan area than to the east side, where absorption has already been attained.

8.5.3 YEAR 2010–2019

The map of 2019 of Delhi shows that the settlement region experienced westward expansion with maximum growth in the western part and comparatively, less growth in the southern part (Figures 8.6 and 8.7). The number of urban areas has grown dramatically, and this growth can be linked to factors including the quickening speed of urbanization and population growth (Tables 8.5 and 8.6). Water bodies are almost dried up except in some eastern parts, vegetation has increased in the southern part, built-up area is now expanding toward the southern part. Open scrub has been decreased and converted into vegetation areas and built-ups.

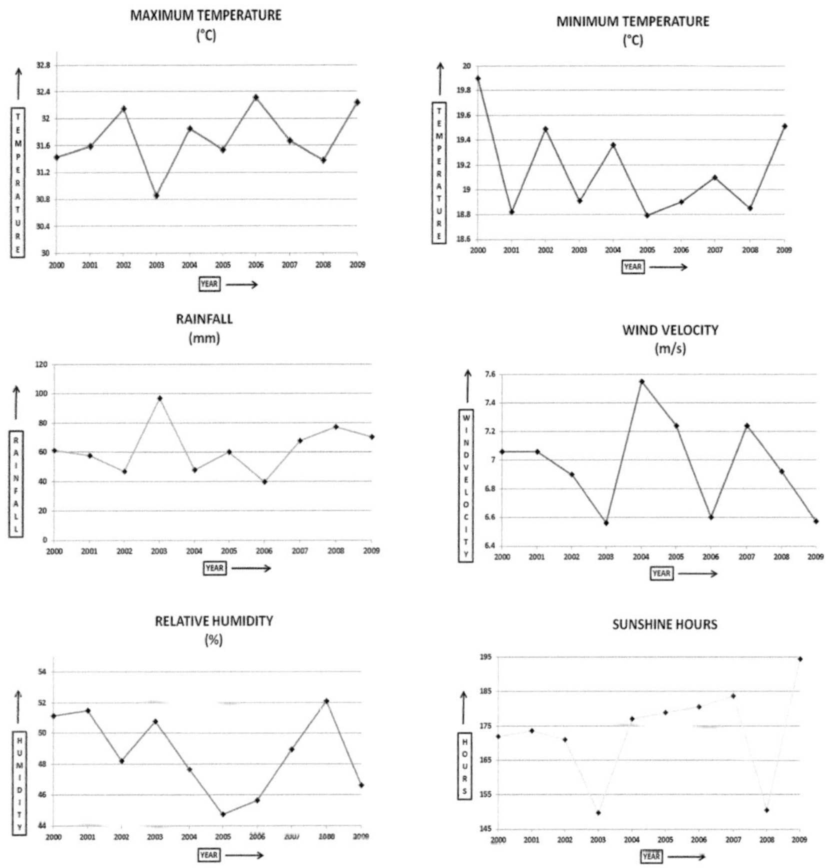

FIGURE 8.4 Graphical representation of climate variability 2000–2009.

8.6 CLIMATE AND LULC CHANGES

8.6.1 FIRST DECADE (1999–2009)

The average temperature in New Delhi fluctuates from 23°C to 40°C. The hottest temperatures are observed in the west, particularly south-west of Delhi. The weather is cool in the city center and south but warm in the eastern and northern regions. The geographic distribution of temperature

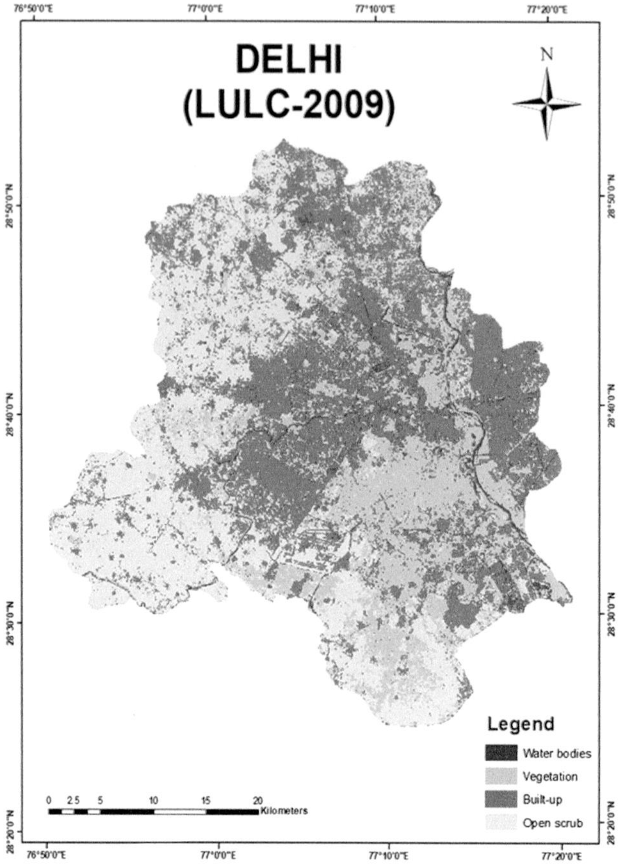

FIGURE 8.5 Land use land cover 2009.

Classification accuracy = 97.5%, Kappa statistics = 0.966%, Kappa statistics percentage = 96.66%

TABLE 8.3 Area for LULC 2009

Classes	Area (sq km)
Water bodies	39.31
Vegetation	398.08
Built-up	458.78
Open scrub	616.80

TABLE 8.4 Confusion Matrix for the Year 2009

	ID	Water Bodies	Vegetation	Built-Up	Open Scrub	Total
Water bodies	1	79	0	1	0	80
Vegetation	2	0	80	0	0	80
Built-up	3	1	0	73	0	74
Open Scrub	4	0	0	6	80	86
Total		80	80	80	80	320

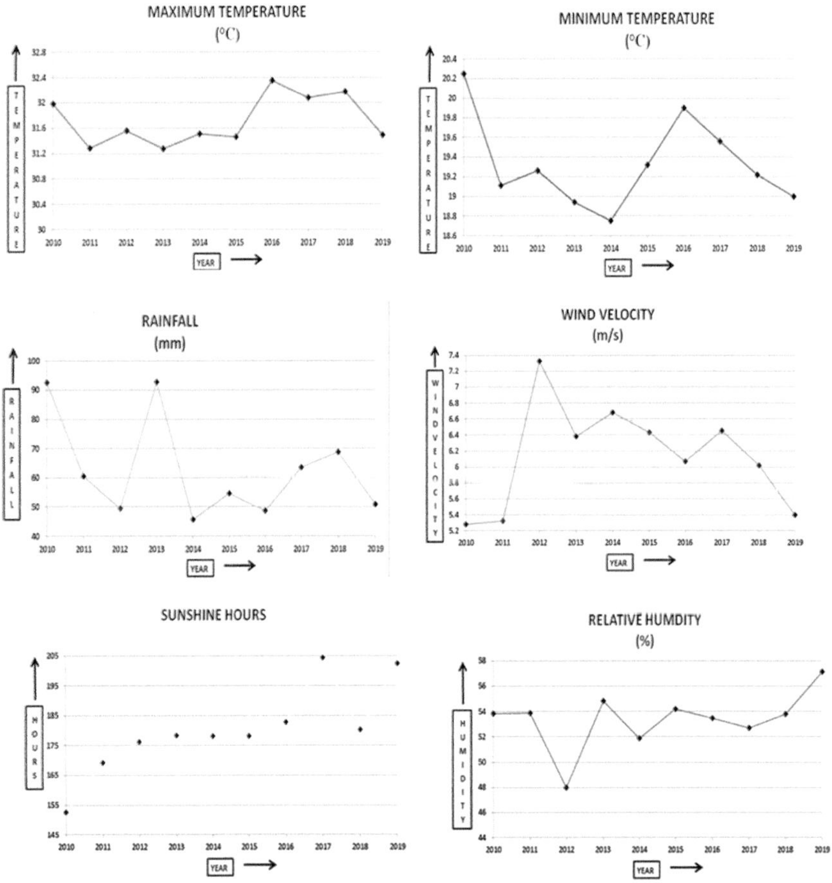

FIGURE 8.6 Graphical representation of climate variability 2010–2019.

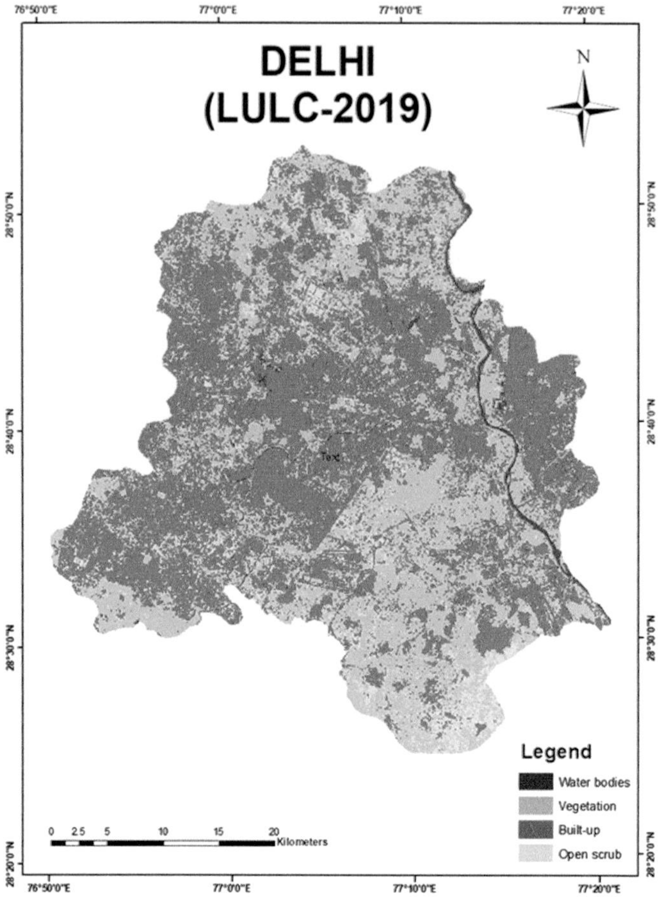

FIGURE 8.7 Land use land cover 2019.

Classification accuracy = 86.56%, Kappa statistics = 0.82%, Kappa statistics percentage = 82.08%

TABLE 8.5 Area for LULC 2019

Classes	Area (sq. km)
Water bodies	28.80
Vegetation	621.47
Built-up	641.96
Open scrub	220.83

TABLE 8.6 Confusion Matrix for the Year 2019

	ID	Water Bodies	Vegetation	Built-Up	Open Scrub	Total
Water bodies	1	75	0	0	0	73
Vegetation	2	4	80	0	38	85
Built-up	3	1	0	80	0	82
Open Scrub	4	0	0	0	42	80
Total		80	80	80	80	320

in Delhi shows that whereas the northeast, center, eastern, southern, and downtown districts are relatively cooler, the more western and southwestern districts exhibit extreme temperatures (Figure 8.8). This splits Delhi city into the following primary temperature regions: extremes of temperature in the west, southwest, and northwest; cold temperatures in the northeast region, center, and southeast; and an average temperature somewhere between the two extreme situations. That is why there is less greenery on the west, southern-west, and northwestern edges of Delhi, and the urban area region is increasing toward these ends, raising the temperature due to overpopulation as well as industrialization. The east portion of Delhi appears to be colder, which is why the Yamuna River is still running and some greenery can be seen. There are variations in the climate variables that are induced by different aspects of land utilization and surface area, and which impact it. While fields of vegetation have a direct connection with decreasing temperatures at the surface, which are responsible for the calming impact in the metro microclimate, cement structures contribute to the already excessive temperatures. The continuous river, Yamuna, has a lowest temperature of 23°C where the amount of water is reported to be highest. The temperature rises to approximately 28°C when the condition of the water's composition changes due to the mingling of substances and particles. The river has a comparatively high capability for heat storing and transferring. Because of the rise in temperature and altered climatic variables, Delhi is also experiencing a lack of rainfall, which explains why there was less greenery throughout the ensuing decade.

8.6.2 SECOND DECADE (2010–2019)

During this time, the Delhi metropolitan landscape has changed dramatically. The environment's precipitation, moisture, and other factors were all affected by LULC in general. Significant findings include an increase in postmonsoon and wintertime temperatures and a decline in the rainy season values. Due to the constant transfer of heat and evaporated water, the water bodies have practically dried out. The open marsh has receded and been replaced by greenery or built-up areas. Although the general temperature in East Delhi rose in some places, it was not as prevalent. This could be due to the Yamuna River's moderating effect. The research area's land areas nearby or near the river Yamuna and other bodies of water, as well as plants, may have contributed to the substantially lower surface temperatures of the metropolitan accumulated region.

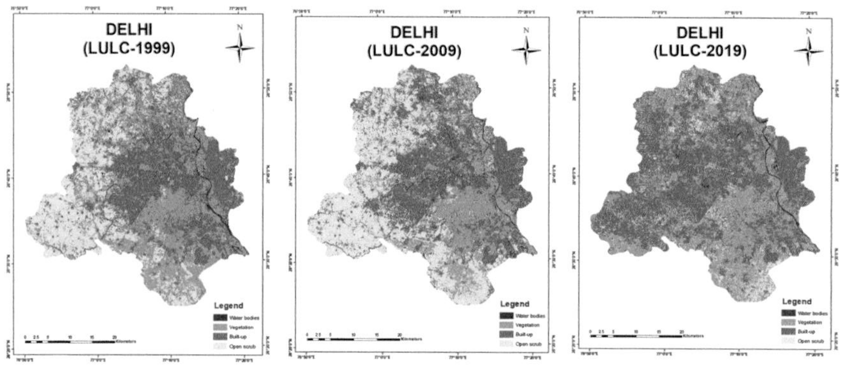

Classes	1999	2009	2019
WATER BODIES	3.59%	2.59%	1.9%
VEGETATION	40.99%	26.31%	41.1%
BUILT-UP	25.37%	30.32%	42.45%
OPEN SCRUB	30.03%	40%	14.06%

FIGURE 8.8 Spatiotemporal analysis for year 1999, 2009.

8.7 CONCLUSION

As a result, the study might result in a comprehensive and coherent spatial-temporal (analytical) analysis of the key bio-physical and climatological procedures influencing land utilization and changing land cover in Delhi. Maps and statistics with helpful choices for alternative land use plans in Delhi were produced with the assistance of land analysis in geospatial data (Janmaijaya et al., 2018; Lambin et al., 2003; Mercy, 2015; Mohan, 2000). A thorough analysis of land use status from 1999 to 2019 reveals significant changes. Although 2019 has seen a significant increase in both built-up area and vegetation; between 1999 and 2009, Delhi's vegetation area shrank as a result of climatic variation, which is partly the result of city's urbanization and industrialization. The microclimatic changes in Delhi are a result of the over-dominance and over-expansion of gray infrastructure. Variability in climatic variables, the urban heat island impact, and the increase of heat waves during the summer are some examples of these microclimatic changes. There are four major classes recognized, with vegetation and built-up areas being the most dominant. Over the course of 20 years, a significant change is noted, the amount of vegetation cover dropped from 40.99% to 26.31% in 2009 and then rose to 41.1% in 2019. The data of settlement area increased from 25.37% in 1999 to 42.25% in 2019. The percentage of water bodies declined from 3.59% in 1999 to 1.9% in 2019. The percentage of open brush, barren land, or land used for ongoing work has decreased from 30.03% in 1999 to 14.06% in 2019. According to Kappa figures, the classification accuracy was 93.41% in 1999, 96.66% in 2009, and 82.08% in 2019. This study clearly shows that LULC are directly correlated with climate, and that as the climate changes, so does the land. For each research year—1999, 2009, and 2019—a map of LULC was created by applying the maximum likelihood approach to the analysis of remote sensing data in a GIS setting. Each study year saw the creation of four separate classes, with a focus on built-up land because it is the result of a variety of human activities and, by definition, affects other classes and contributes to climate change. The results showed that the fear of global warming caused a rapid increase in the amount of vegetation and built-up area. The Indian government has implemented numerous climate-related policies and initiatives. Furthermore, given that the projection results showed that arable land and built-up areas would grow quickly and that open space and water bodies would have substantially decreased by 2030, it was noted that changes by that year would probably follow the trend shown between 1999 and 2019. These will have a significant impact on the city's local climate and increase greenhouse gas emissions since they would have severely reduced carbon sinks and caused unpredictability in temperature, precipitation, wind speed, humidity, and other factors. Better urban design and rapid afforestation are required for this project, taking into account how agriculture helps to mitigate the effects of climate change and global warming. The conflicting land uses that the LULC changes showed, especially those involving vegetation, have demonstrated that there has been a significant loss of vegetation in Delhi's first 10 years (1999–2009), which has accelerated climate change. The decline in greenery resulted in a net variation in high temperatures and a minor increase in minimum yearly temperatures when comparing the changes to trends in precipitation and temperature. The rainfall trend showed how extreme weather had a detrimental effect on the vegetation. However, in the second decade, which runs from 2009 to 2019, people began planting trees in the open scrub area due to concerns about global warming, which ultimately resulted in a decrease in the amount of open scrubland. Urban development is unavoidable, but it may be controlled and directed in a way that preserves arable land and is desired and sustainable with the right management and planning. It is anticipated that the study would significantly contribute to the development of sound change detection approaches and improved policy-making for India's National Capital Region. Therefore, to reduce the effects of microclimatic changes, Delhi's national capital region needs to be redesigned and efficiently planned. Green roofing, roadside vegetation, trackside vegetation, and urban green spaces are some regulations that, when properly and effectively implemented, will have a beneficial effect.

KEYWORDS

- **land use/land cover**
- **climate changes**
- **surface temperature**
- **water bodies**
- **vegetation**

REFERENCES

Budhiraja, B., Pathak, P., and Agrawal, G. (2017). Spatial-temporal variability of urban heat islands in local climate zones of Delhi NCR. In: *Conference Paper.* October 2017. https://doi.org/10.1117/12.2280253

https://doi.org/10.1146/annurev.energy.28.050302.105459
https://www.britannica.com/place/Delhi
https://www.epa.gov/scram/meteorological-data-and-processors

Balogun, I. A., Adeyewa, D. Z., Balogun, A. A., and Morakinyo, T. E. (2011). Analysis of urban expansion and land use changes in Akure, Nigeria, using remote sensing and geographic information system (GIS) techniques. *Journal of Geography and Regional Planning*, *4*(9), 533–541.

Chen, X.-L., Zhao, H.-M., Li, P.-X., and Yih, Z.-Y. (2005). Remote sensing image-based analysis of the relationship between urban heat island and land use/cover changes. *Remote Sensing of Environment 104*(2006) 133–146. https://doi.org/10.1016/j.rse.2005.11.016

Hageback, J., Sundberg, J., Ostwald, M., Chang, D., Yun, X., and Knutsson, P. (2005). Climate variability and land use change in Danangou Watershed, China—examples of small scale Farmer's adaptation. *An Interdisciplinary, International Journal Devoted to the Description, Causes, and Implication of Climate Change 72*(1), 189–212. https://doi.org/https://doi.org/10.1007/s10584-005-5384-7

Hassan, Z., Shabbir, R., Ahmad, S. S., Malik, A. H., Aziz, N., Butt, A., and Erum, S. (2016). Dynamics of land use and land cover change (LULCC) using geospatial techniques: A case study of Islamabad Pakistan. *SpringerPlus 5*, 812. https://doi.org/10.1186/s40064-016-2414-z

Janmaijaya, M., Singh1, R. B., and Dhaka, S. K. (2018). Detection of temperature variability and trends in the lower troposphere over Delhi: A study of joint influence of ENSO and Land use/Land cover during 1980–2015. *Singapore Journal of Tropical Geography 40*(2019) 111–123. https://doi.org/10.1111/sjtg.12251

Lambin, E. F., Geist, H. J., and Lapers, E. (2003). Dynamics of land use and land cover change in Tropical region. *Annual Review of Environment and Resources 28*, 205–241.

Mercy, M. W. (2015). Assessment of the effects of climate change on land use and land cover using Remote sensing: A case study from Kenya. *Dresden Nexus Conference Working Paper Series, DNC2015/03*.

Mohan, M. (2000). Climate change: Evaluation of ecological restoration of Delhi ridge using remote sensing and GIS Technologies. *International Archives of Photogrammetry and Remote Sensing XXXIII*, Part B7. Amsterdam 2000.

Mridha, N., Chakraborty, D., Roy, A., and Kharia, S. K. (2012). *Role of Remote Sensing in Land Use and Land Cover Modelling*. New Delhi: Indian Agricultural Research Institute.

Parveen, S., and Sarkar, S. (2018). Land use land cover mapping with change detection: A spatio-temporal analysis of NCT of Delhi from 1981 to 2015. *International Journal of Geomatics and Geosciences 8*(1), 41–54.

Roy, S. S. (2011). An analysis of local spatial temperature patterns in the Delhi metropolitian area. *Physical Geography 32*(2), 114–138 https://doi.org/10.2747/0272-3646.32.2.114

CHAPTER 9

MEASUREMENTS OF PRECIPITATION OVER THE ANTARCTIC REGION: GROUND-BASED DATA AND SATELLITE MEASUREMENTS

STEPHY ABRAHAM[1], ALIA YUSUF[2], ROUNAQ GOENKA[3], and ZIAUR REHMAN ANSARI[1]

[1]*Interdisciplinary Department of Remote Sensing & GIS Applications, Aligarh Muslim University, Aligarh, Uttar Pradesh, India*

[2]*Geology Section, Women's College, Aligarh Muslim University, Aligarh, Uttar Pradesh, India*

[3]*National Atmospheric Research Laboratory, National Remote Sensing Centre, Department of ISRO, Government of India, Balanagar, Hyderabad, India*

ABSTRACT

Precipitation in Antarctica is not very common but it plays an essential role in the context of global warming, melting, and sea-level rise. Moreover, precipitation in the form of rain or snow has a significant contribution toward the surface mass balance of ice sheets in the Antarctic region, which has a direct impact on sea level rise. Sensors onboard satellites are currently the only instruments that can provide global precipitation measurement. However, the satellite data needs to be validated before being used for studies mentioned above. As part of Indian expeditions to Antarctica, a ground-based Micro Rain Radar instrument during the

Sustainable Development: A Geospatial and Statistical Perspective. Qazi Mazhar Ali, Rizwan Ahmad, Irfan Ali, & Prasenjit Chatterjee (Eds.)

DOI: 10.1201/9781779643247-9

Antarctic 2017/2018 campaigns provided unprecedented precipitation data over the Southern Ocean. This instrument provided a continuous vertical structure of precipitation, rain rate, and radar reflectivity. In this study, the results of the comparison between ground-based data and satellite-retrieved precipitation products have been presented. The study reviews the contribution of satellite observation for better precipitation estimates over the continent.

9.1 INTRODUCTION

Precipitation is formed from the condensation of water vapor in the atmosphere. There are different forms of precipitation including drizzle, sleet, rain, snow, ice, pellet, graupel, and hail. Precipitation occurs with the saturation of water vapor and its condensation, which when becomes heavy enough falls under gravity as precipitation. It is an essential element of the water cycle, which results in depositing the major amount of fresh water on earth. It also furnishes a favorable environment for a variety of ecosystems, crop irrigation, and hydroelectricity generation. Therefore, precipitation is an important parameter in the global climate system. Studying precipitation is imperative for understanding its contribution to the water cycle, and improvement in its measurement is also an important factor to achieve progress in our understanding of the hydrological cycle (Tang et al., 2014). With changing climate, there is a change in spatial and temporal distribution of precipitation and there are studies indicating an increase in extreme rainfall events, which makes monitoring and measurement of precipitation data very important.

Satellite rainfall estimates in conjugation with ground-based observation are widely used to supplement gauge observations. However, validation is necessary for building confidence in using the satellite products for various applications. Some global and regional validations have been reported by different workers for different satellite rain products (Dinku et al., 2007; Feidas, 2010).

9.2 PRECIPITATION IN ANTARCTICA

Precipitation in Antarctica is very much like that of a desert region. It is lower than 50 mm per year in the interior, while near the coasts, it exceeds

300 mm, with the highest rainfall of 600 mm. Genthon et al. (2009) analyzed Antarctic precipitation and climate change using different modules. Since cold air cannot hold much moisture, it does not produce a heavy amount of precipitation and hence falls under among some of the most arid regions in the world. At the South Pole, the total precipitation amounts to nearly 2 mm per year, so it practically never snows or at most, a very light snow falls. David et al. (2013) evaluate the temporal variability of the Antarctic surface mass balance, approximated as precipitation minus evaporation (P−E) and Southern Ocean precipitation in five global reanalyses during 1989–2009. The average precipitation at the Casey Station, which is on the coast, receives a higher amount of precipitation, largely in the form of snow.

9.3 STUDY AREA

The study area is the earth's farthest southern continent which contains the geographic South Pole. It is situated within the Antarctic region of the southern hemisphere, almost completely south of the Antarctic Circle, and is encompassed by the Southern Ocean. About 98% of Antarctica has an iced roof that averages 1.9 km in thickness, which extends to the entire region, but the northernmost reaches of the Antarctic Peninsula. Antarctica, on average is the coldest, driest, and windiest continent and has the very best average elevation of all the continents.

9.4 GROUND DATA USED

For the present study, ground-based radar and satellite precipitation measurement sensor is used. Ground measurement of precipitation is very challenging in some extreme climate areas like the poleward Antarctic. An expedition to coastal Antarctica as part of the 10th Southern Ocean Expedition (SOE) of India was conducted by the National Centre for Polar and Ocean Research, Goa to survey Southern Ocean waters. The ship expedition carried the Micro Rain Radar (MRR) instrument for the measurement of precipitation in coastal Antarctica. The measurement was carried out completely in the ocean area.

Ground-based radar is used to generate rainfall data products. Radar-based quantitative rainfall estimation is obtained at certain altitudes, which depends on radar elevation angle and distance from the radar. The radar elevation angle is measured as an angle between the horizontal plane and the line of sight which is measured in the vertical plane. Vertically pointing RADAR MRR along the German Baltic coast was used for measuring the vertical variability of rainfall and rainfall size distribution variability (Peters et al., 2002).

9.5 MRR

Souverijns et al. (2018) compared CloudSat snowfall products against MRR over the Antarctic region. The measurement of profiles where extremely minute amounts of precipitation below the threshold of traditional rain gauges are discernible is done using MRR, which uses 24-GHz frequency radar with great sensitivity and precise temporal resolution. Following its propagation back via precipitation medium, the broadcast waveform is received by the radar after being scattered back from the particles in the resolution volume (Chandrasekar et al., 2008).

An integrated mixing diode and frequency-regulated Gunn-diode oscillator constitute the fundamental part of the radar. Fifty megawatts (MW) is the nominal transmission power. A horn and waveguide serve as the feed for an offset paraboloid dish with a diameter of 60 cm, which receives the linearly polarized radio frequency power. Using the same antenna assembly, the backscattered signal is received (monostatic radar). Mounted in the wave path between the horn and the Gunn oscillator, a mixing diode detects the received signal. As a mixer, this diode is biased with a portion of the broadcast signal. Because the receiver also stops working when the transmitter shuts off, this straightforward design cannot be used in pulsed mode. When the diode is run in continuous wave mode, a voltage that is utilized for additional signal processing and is dependent on the phase difference between the transmit and receiving signals emerges at the output. It gauges the distributions of drop sizes that are obtained from rainfall factors, including precipitation rates, liquid water content, and distinctive falling velocity.

The compact, lightweight, and simple-to-use MRR is a novel and inventive device. For the time being, it can be used to cast precipitation, that is, it can detect when rain starts to fall from far above the radar a few minutes before it starts to fall at ground level. It is a very dependable system that may be used in harsh and isolated settings. It requires little upkeep and is primarily designed for extended periods of unsupervised operation. The data may be accessed in several formats, including raw, averaged, and latest data sets. Controlling the settings, visualizing them, and exporting them into different forms for use in other programs for reading, publishing, storing, and other purposes is made possible by the accompanying software package.

Continuous waveform transmission mode is the RADAR's output signal. Selectable range resolution profile measurements may be obtained by using frequency modulation mode, which modulates the transmit signal with a linearly diminishing saw tooth.

9.6 SATELLITE DATASETS USED

Combined data from passive microwave sensor in low earth orbit and infrared radiometer in geostationary orbit provide high-resolution rainfall products (Kubota et al., 2009).

For the study, mainly several datasets from Defense Meteorological Satellite programs were used. Very-high-resolution visual and infrared spectrum imagery can be used either day or night. The primary data sensor used in the DMSP is a scanning radiometer, which provides signals for four separate types of pictorial imagery. The high-resolution visual channel provides imagery useful for meteorological purposes with scene illumination levels varying from subsolar to that obtained from one-fourth moon conditions (Spangler, 1974).

- **DMSP-F16** was launched on October 18, 2003 and orbiting at an altitude of 848 km. The main mission of F16 is cloud imaging and MW imaging/temperature/humidity sounding.
- **DMSP-F17** was launched on November 4, 2006 and orbiting at an altitude of 848 km, which is similar to DMSP F16. The main mission of this satellite program is also similar to DMSP F16.
- **DMSP-F18** was launched on October 18, 2009 and orbiting at an altitude of 850 km. The main mission of this satellite program is also similar to DMSP F16 as well as DMSPF17.
- **DMSP F16, F17,** and **F18** have a Sun synchronous orbit. The main mission of this satellite program was to cloud imaginary and MW imaging/temperature/humidity sounding and made some substantial contribution to SpaceWeather.

Sensors used in DMSP F16, F17, and F18 are:

- OLS (Operational Line Scan System),
- SSMIS (Special Sensor Microwave–Imager/Sounder),
- SESS/SSUSI SESS/Special Ultraviolet Spectrographic Imager,
- SESS/SSULI SESS/Special Ultraviolet Limb Imager,
- SESS/ SSI/ ES-3SESS/Special Sensor Ionosphere Plasma Drift/Scintillation Monitor-3, SESS/SSM-Boom SESS/Special Sensor Magnetometer—Boom,
- SESS/SSJ5 SESS/Special Sensor Precipitating Electron and Ion Spectrometer

9.7 SOFTWARE USED

9.7.1 MATLAB

MATLAB (matrix laboratory), a fourth-generation high-level programming language, was developed by Math Works, which is used for technical computing. It integrates numerical computation, visualization, and programming within an easy-to-use environment where problems and solutions are expressed in familiar mathematical notations. MATLAB is developed by Math Works. It is an interactive system that does not require dimensioning, whose basic data element is an array.

Image Processing Toolbox provides a comprehensive set of reference-standard algorithms, functions, and apps for image processing, analysis, visualization, and algorithm development.

Image analysis, image segmentation, image enhancement, noise reduction, geometric transformations, and image registration can be easily performed in MATLAB. It also allows matrix manipulations; plotting of functions and data; implementation of algorithms; creation of user interfaces; interfacing with programs written in other languages, including C, C++, Java, and FORTRAN; analyzes data; develop algorithms; and make models.

9.8 VALIDATION METHOD

Table 9.1 shows the rain duration during SOE where satellite passes were available. As we can see, rainfall events were both short duration and long duration. MRR was configured to collect rainfall data every 30-s time resolution.

TABLE 9.1 Rain Duration during SOE Where Satellite Passes Were Available

Date	Time Range	Satellite Passes	Remarks
08/12/2017	1300 to 1430 UTC	f17des	Ship stationary
11/12/2017	1630 to 1830 UTC	amsr2des, f17des	Drizzle
11/12/2017	1730 to 2200 UTC	amsr2des,	
14/12/2017	1610 to 17:35 UTC	f18des, f17des	Drizzle, light rainfall
14/12/2017	1735 to 0000 UTC	amsr2des, f16asc	Constant rain, bad weather
15/12/2017	0000 to 0650 UTC	f17asc	Constant rain, bad weather
15/12/2017	1010 to 1020 UTC	amsr2asc	Drizzle
17/12/2017	0225 to 0235 UTC	f18des, f18des	Light rain
17/12/2017	2130 to 2140 UTC	amsr2des	
18/12/2017	0207 to 0253 UTC	f18asc	
18/12/2017	2240 to 2250 UTC	f16asc	Drizzle
19/12/2017	0925 to 0945 UTC	amsr2asc	
20/12/2017	0920 to 2000 UTC	amsr2des, f18des, f17des, f16des, amsr2asc	
20/12/2017	2030 to 2240 UTC	f16asc	
21/12/2017	0150 to 0200 UTC	f18asc	
21/12/2017	1310 to 1840 UTC	f18des, f17des	Snow/rain
22/12/2017	0030 to 0110 UTC	f18asc, f17asc	
22/12/2017	1425 to 1440 UTC	f17des,	
23/12/2017	0350 to 1300 UTC	f16des, amsr2asc	Multiple events
24/12/2017	1310 to 1540 UTC	f18des, f17des,	
25/12/2017	0040 to 0115 UTC	f18asc	Very small events
25/12/2017	2100 to 2300 UTC	f16asc	Scattered individual events
26/12/2017	0900 to 1730 UTC	f18des, f17des, f16des, amsr2asc	
17/01/2018	1100 to 1210 UTC	amsr2asc	
17/01/2018	1610 to 1620 UTC	f18des	
17/01/2018	1920 to 0000 UTC	amsr2des, f16asc	
18/01/2018	0000 to 0240 UTC	f18asc, f17asc	
18/01/2018	1120 to 1130 UTC	amsr2asc	
18/01/2018	2020 to 0000 UTC	amsr2des, f16asc	
19/01/2018	0000 to 0350 UTC	f18asc, f17asc	
19/01/2018	1235 to 1530 UTC	f16des,	
19/01/2018	1715 to 2359 UTC	amsr2des, f18des, f16asc	
20/01/2018	0000 to 0150 UTC	f18asc, f17asc	
20/01/2018	1600 to 1640 UTC	f17des	
21/01/2018	0000 to 0310 UTC	f18asc, f17asc	
21/01/2018	1950 to 2359 UTC	amsr2des, f16asc	

Figures 9.1–9.4 show the MRR graphics software interface. Here, we can see how the data can be visualized as a function of both time and height. Color represents a range of values for the physical parameter being visualized. The scale is displayed at the bottom right of the screen. MRR does not have any in-built global positioning system (GPS); we rely on the ship's GPS data. Then, we manually correlate the data collected to the ship's GPS coordinates so that satellite validation can be performed according to the ship's location and corresponding satellite pixel.

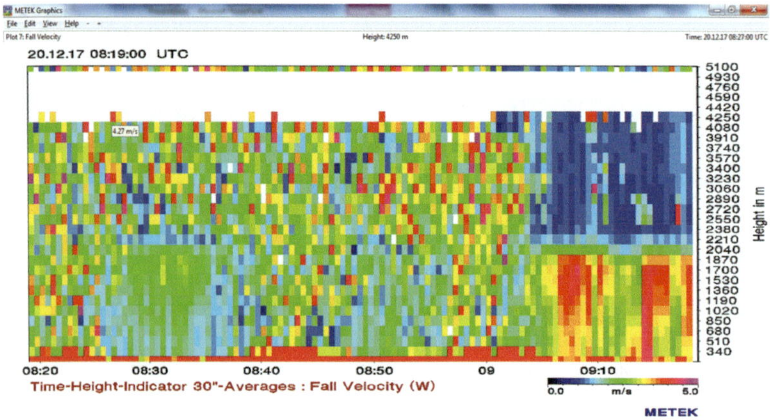

FIGURE 9.1 The time height indicator of average fall velocity (W) of Antarctic precipitation by MRR on December 20, 2017.

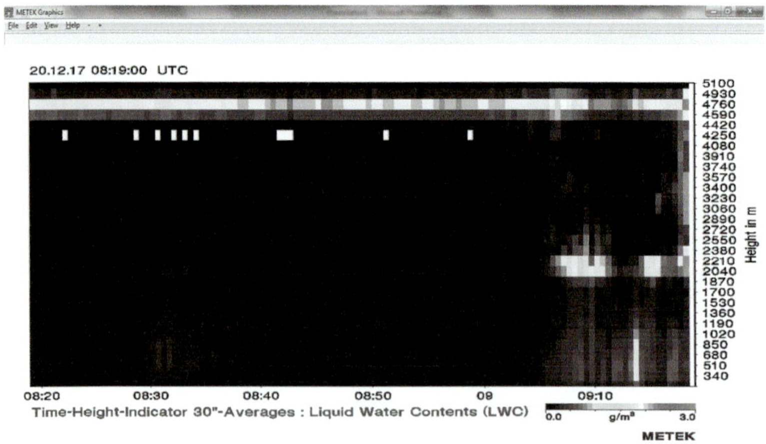

FIGURE 9.2 The time height indicator of liquid water content of the Antarctic precipitation measured by MRR on December 20, 2017.

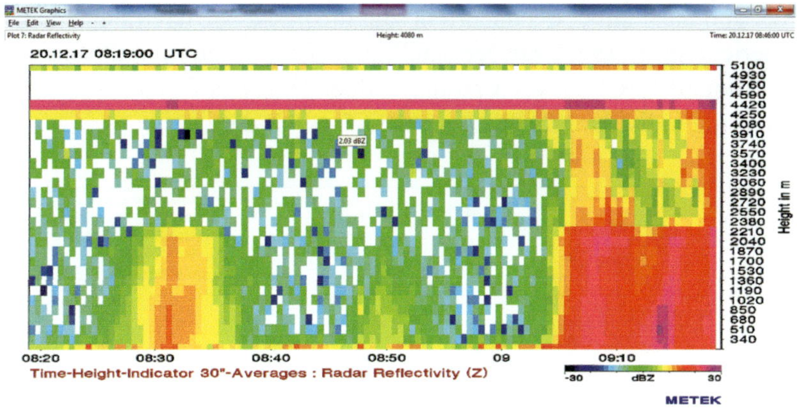

FIGURE 9.3 The time height indicator of radar reflectivity (Z) measured by MRR on December 20, 2017.

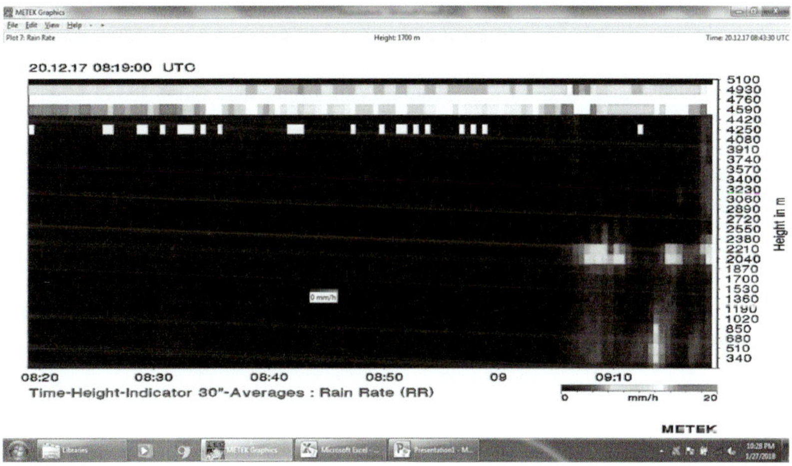

FIGURE 9.4 The time height indicator of rain rate of Antarctic precipitation by MRR on December 20, 2017.

9.9 RESULT

Figures 9.5 and 9.6 show time series plots of the rainfall data collected during SOE-10. The data has been segregated in between 0 and 1 and above 1 mm/h because we have a significant number of points range between 0 and 1 mm/h due to the high sensitivity of the MRR instrument. These figures demonstrate the number of events that we recorded for

MRR and the number of concurrent satellite passes we have along with the numeric values. As we can see satellite data shows no rainfall sometimes even though MRR shows a rainfall value. This was further analyzed by comparing individual satellite passes with MRR data.

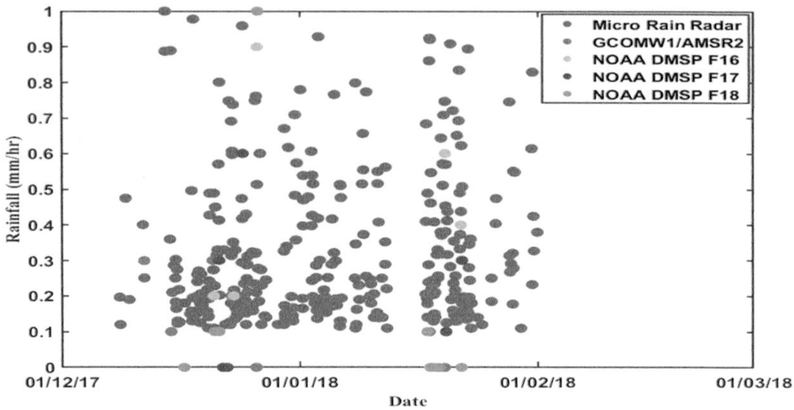

FIGURE 9.5 Time series of available rainfall data from MRR and different satellites for the rain rate range of 0–1 mm/h.

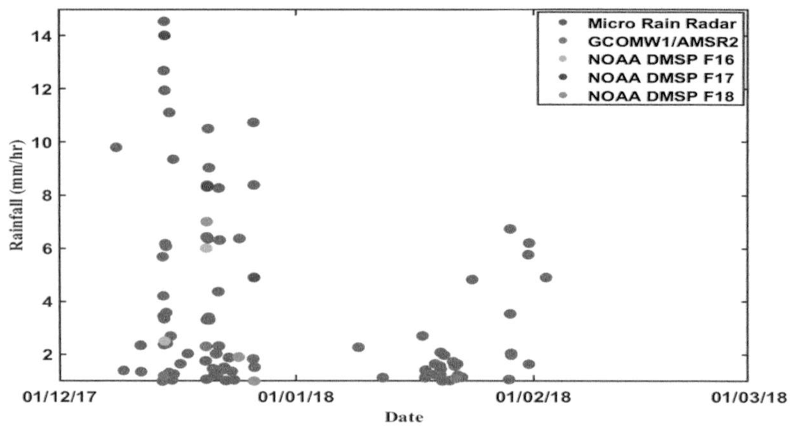

FIGURE 9.6 Time series of available rainfall data from MRR and different satellites for the rain rate range greater than 1 mm.

Figure 9.7 shows the comparison of individual satellite data with the MRR instrument. We have segregated data between 0 and 1 and above 1 mm/h separately due to large number of points in 1 mm/h window. Also, we have compared this rainfall data with MRR surface rainfall and average for the first three heights, that is, 0–600 m. The correlation coefficient values are displayed in Table 9.2. As we can see the comparison is best for surface rainfall with considerable differences in average rainfall. Also, if we see individual satellites SSMIS sensors show a good relation with MRR surface rainfall compared to the AMSR2 sensor. The overall correlation for surface rainfall was 0.88 for surface rainfall and 0.19 for average rainfall.

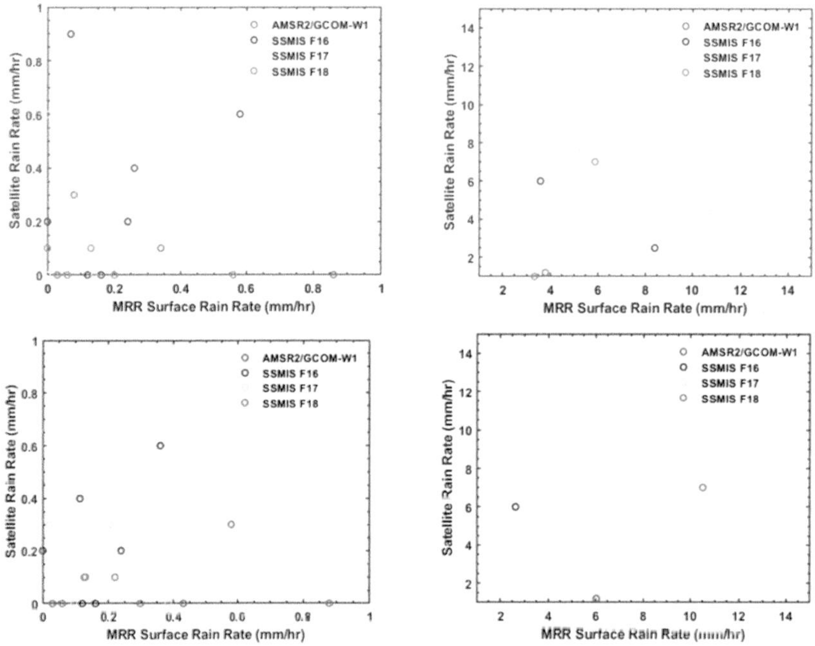

FIGURE 9.7 Validation of satellite data with MRR instrument for both surface and average rainfall (0–600 m) for 0–1 mm/h and greater than 1 mm/h.

TABLE 9.2 Correlation Coefficient for Individual Satellites with MRR Surface Rainfall and Height Average Rainfall

Satellite	MRR Surface	MRR Average
AMSR2	0.22	0.02
SSMIS F16	0.62	0.12
SSMIS F17	0.96	0.68
SSMIS F17	0.86	0.006

KEYWORDS

- **Antarctica**
- **Micro Rain Radar**
- **precipitation**
- **satellites**

REFERENCES

Chandrasekar, V., Hou, A., Smith, E., Bringi, V. N., Rutledge, S., Gorgucci, E., Petersen, W., and Jackson, G. (2008). Potential role of DualPolarization radar in the validation of satellite precipitation measurements: Rationale and opportunities. *Bulletin of the American Meteorological Society 89*(8), 1127–1146. https://doi.org/10.1175/2008BAMS2177.1.

David, H. B., Julien, P. N., and Andrew, J. M. 2013 Supplementary information for: Corrigendum: Central West Antarctica among the most rapidly warming regions on Earth, January 2014. *Nature Geoscience 7*(1), 76. https://doi.org/10.1038/ngeo2016

Dinku, T., Ceccato, P., Grover-Kopec, E. K., Lemma, M., Connor, S. J., and Ropelewski, C. F. (2007). Validation of satellite rainfall products over East Africa's complex topography. *International Journal of Remote Sensing 28*, 1503–1526. https://doi.org/10.1080/01431 160600954688

Feidas, H. (2010). Validation of satellite rainfall products over Greece. *Theoretical and Applied Climatology 99*, 193–216. https://doi.org/10.1007/s00704-009-0135-8.

Genthon, C., Krinner, G., and Castebrunet, H. (2009). Antarctic precipitation and climate-change predictions: Horizontal resolution and margin vs plateau issues. *Annals of Glaciology 50*, 55–60. https://doi.org/10.3189/172756409787769681.

Kubota, T., Ushio, T., Shige, S., Kida, S., Kachi, M., Okamoto, K. (2009). Verification of high-resolution satellite-based rainfall estimates around japan using a gauge-calibrated ground-radar dataset. *Journal of the Meteorological Society of Japan 87A*, 203–222. https://doi.org/10.2151/jmsj.87A.203.

Peters, G., Fischer, B., and Andersson, T. (2002). Rain observations with a vertically looking Micro Rain Radar (MRR). *Boreal Environment Research 7*, 353–362.

Souverijns, N., Gossart, A., Lhermitte, S., Gorodetskaya, I. V., Grazioli, J., Berne, A., Duran-Alarcon, C., Boudevillain, B., Genthon, C., Scarchilli, C., and van Lipzig, N. P. M. (2018). Evaluation of the CloudSat surface snowfall product over Antarctica using ground-based precipitation radars. *The Cryosphere 12*, 3775–3789. https://doi.org/10.5194/tc-12-3775.

Spangler, M. J. (1974). The DMSP primary data sensor. In: *Conference on Aerospace and Aeronautical Meteorology, 6th, El Paso, Tex., November 12-15, 1974, Preprints. (A75-35351 16-47)* Boston, Mass.: American Meteorological Society, 1974, pp. 150–157.

Tang, L., Tian, Y., and Lin, X. (2014). Validation of precipitation retrievals over land from satellite-based passive microwave sensors. *Journal of Geophysical Research: Atmospheres 119*(8), 4546–4567.

CHAPTER 10

SPATIOTEMPORAL MONITORING OF LAND SURFACE TEMPERATURE FOR ASSESSING HEAT-RELATED HEALTH RISKS

ZEBA KHANAM[1], MANSOOR ALAM SIDDIQUI[2], UZMA AJMAL[3], SALEHA JAMAL[3], and S. K. AJIM ALI[3]

[1]*Post Doctoral Fellow (ICSSR), Jawaharlal Nehru University, New Delhi*

[2]*Department of Geography, Women's College, Aligarh Muslim University, Aligarh, Uttar Pradesh, India*

[3]*Department of Geography, Aligarh Muslim University, Aligarh, Uttar Pradesh, India*

ABSTRACT

Urban heat island (UHI) phenomena are increasingly recognized as the most evident characteristics of urban climate. The intensity and scale of UHI are increasing in urban areas with ongoing urbanization. This chapter analyzed the Bulandshahr city, India, within a time frame of 16 years (2003–2018) using satellite images of Landsat 7 and 8 for 2003 and 2018. With the help of RS data, land surface temperature (LST), normalized difference vegetation index (NDVI), and normalized difference built-up index (NDBI) were calculated. In this study, LST was used and correlation was estimated to study the effect of green cover and built-up construction on LST using NDBI, NDVI, and LST. Second, this study explores the impact of heat island on self-reported human health, that is, sleep deprivation and heat stress, using primary data of different city wards. This study also concludes

Sustainable Development: A Geospatial and Statistical Perspective. Qazi Mazhar Ali, Rizwan Ahmad, Irfan Ali, & Prasenjit Chatterjee (Eds.)

DOI: 10.1201/9781779643247-10

that health risk in urban population had a disproportionate impact on more vulnerable social groups. Furthermore, this study summarizes that the impact of the UHI on heat-related health risk is significant, a frequently underestimated factor. This is especially true when analyzing smaller cities such as Bulandshahr city.

10.1 INTRODUCTION

The world is getting urbanized at a very fast rate, and this phenomenon is more specific in third-world countries. The rapid surge in urbanization, urban population, and spreading out of urban land has resulted in the extension of the built-up area and reduction of the green spaces (Shahmohamadi et al., 2011). Urban lands are expanding at a speedy rate owing to the increasing demand of land for various residential and other activities. With urbanization and extension of urban lands, green spaces are being reduced and turning into built-up surfaces. Previous research works have proved that urbanization can ominously affect the regional weather and climate (Singh et al., 2017). Studies have concluded that due to decreases in green areas and extension of built-up surfaces, the land surface temperature (LST) of an area is expected to be increased. The rise in surface temperature can be attributed to human activities such as urban development, changes in land use for different purposes that disrupt regional climate, alterations in air movement, shifts in land cover, and adjustments in surface energy distribution. These factors have led to changes in emission patterns, impacting heat transfer processes. The expansion of built-up surfaces and reduction of vegetated areas are recognized as crucial contributors to the rise in land temperatures.

Various research works are performed to explore the influence of increasing urbanization on land surface temperatures. These research works have established that speedy urbanization is leading to shrinking vegetation, elevated surface temperatures and change in urban micro-climate. The increasing land surface temperatures have also affected the population in many ways. Research works have been carried out worldwide on influence of increasing land temperatures and its probable effect on urban susceptibility. The rise in land temperature has significant effects including increased energy consumption, disruption of ecosystems, impacts on public health, and alterations in local climate patterns.

The land 7 ETM and 8 Operation Land Imager (OLI) are widely used to derive the vegetation cover, built-up areas, and the land surface temperatures. The thermal infrared bands of the Landsat imagery are able to derive the surface temperature. The information of radiant surface temperature is important to a variety of issues and interests in the urban microclimate change, global environmental change, and man environment relations (Liu and Zhang, 2011). Earlier studies have explicitly documented that diminishing vegetated areas and up surging built-up areas are repercussion of rapid urban growth that have altered the urban microclimate.

According to Intergovernmental Panel on Climate change, human-induced changes in microclimate are showing a rising trend all over the globe. Furthermore, the 5th report of IPCC indicates an increase in frequency, length, and intensity in heat waves all over the land, in near future. Many studies have identified the detrimental effect of heat wave on human health across the globe (Anderson and Bell, 2011; Loughnan et al., 2012; Michelozzi et al., 2009; Sung et al., 2013; Zeng et al., 2016). These studies also demonstrated that heat related phenomena can cause illness among the residents of an area irrespective of absolute temperature. Although all segments of population are vulnerable to heat related causalities but some particular groups are more vulnerable than others. People living in low-income, densely populated areas with limited access to urban green spaces are considered more vulnerable within the population (Bambrick, 2011). Human activities that contribute to heat in urban areas can impact the mortality rates among vulnerable populations (Tomlinson et al., 2011).

Since, temperature is increasing in urban local areas, it may result into Urban Heat Island (UHI) phenomena which further may affect the microclimate of an area as well as the number of hot days. Similarly the potential risk of heat stress and sleep disturbance can increase within the urban locals (Tan et al., 2010; Shahmohamadi et al., 2011). UHI not only increases the risk of potential health threat, but it can also deteriorate the quality of life of the residents (Tan et al., 2010). Bulandshahr city diverges from typical tropical climates, as it encounters a subtropical climate characterized by mild, dry winters and hot, rainy summers. This climatic pattern increases the risk of heat-related fatalities. During the month of April and May, a dry and oppressing local wind blows in the northern parts of India from Punjab to Bihar with high intensity between Delhi and Patna due to intense heating of ground surface temperature. In local terms, this wind is called loo, which can increase the heat-related causalities in an area. Simultaneously human activities are increasing UHI phenomena which further can increase the possibility of heat waves in urban local areas. Most of the studies that are related to heat mortality have projected health threats as consequence of excessive heating of the surface (Taha, 1996; Rosenfeld et al., 1998; Akbari et al., 1997). The direct impacts of the changing climate on health is reported by exposure to augmented temperature especially during the time of heat waves (Collier, 2006; Tan et al., 2010; Hester and Bauman, 2013).

This study made an attempt to analyze the microclimate of a small city using the indices such as normalized difference vegetation index (NDVI), normalized difference built-up index (NDBI), and LST to study the heat island phenomena and correlations between each variable. The study calculated LST to understand the heat island phenomenon, utilized NDVI to assess vegetation cover, and analyzed NDBI to study built-up areas in the study area using Landsat 7 and Landsat 8 images for the years 2003 and 2018. After considering the correlation of each variable, researcher made a further attempt to explore the impact of heat island on human health. Health data was collected through questionnaire survey for the year 2018. Time-based spatial evaluation and temperature estimation were done using remotely sensed data, ArcGIS and ERDAS software, whereas Karl Pearson's coefficient of correlation was used to study correlation between the variable such as NDVI, NDBI, and LST. This chapter is a microlevel study of a small city, that is, Bulandshahr. Bulandshahr city is facing the challenges of unplanned and unchecked growth from last few decades and the impact of deteriorating environment is visible in the city. Hence, the present study asserts that it is the sole research endeavor investigating the correlation between microclimate and health in Bulandshahr city, with the aim of assessing the influence of urbanization on smaller urban areas.

10.2 STUDY AREA

Bulandshahr city is a district headquarter of Bulandshahr district which is located $28^0 4'$ N of Latitude and $77^0 52'$ E of latitude. Bulandshahr city is situated at 225.5 m above mean sea level with total area is approximately 18 km^2. Bulandshahr city is situated in plain soil alluvial fertile land of Ganga and the newer alluvial plain of river Kali flowing from northwest to southeast of city. Bulandshahr city is located at a distance of 68km^2 to the Delhi, the National Capital of India. Having situated in the National Capital Region (NCR) of India, it has experienced the rapid urbanization and unexpected population growth in last few decades (Shriram, 2017). The trend of urbanization in NCR region is much higher than the national average (Lokhande et al., 2015), due to its location in the close vicinity of the NCR. Therefore, the urban landscape of Bulandshahr city has undergone change from time to time. During primary survey, it was observed that rapid urbanization has negatively affected the quality and comfort of urban population of the study area. This city as a center of Bulandshahr district has experienced rapid pace of population growth over the last three decades from 127,201 people in 1991 to more than 230,024 in 2018. Analyzing these trends and employing an advanced approach to study microclimate and its impact on human health can provide valuable information for policy makers and urban planners for future endeavor (Figure 10.1).

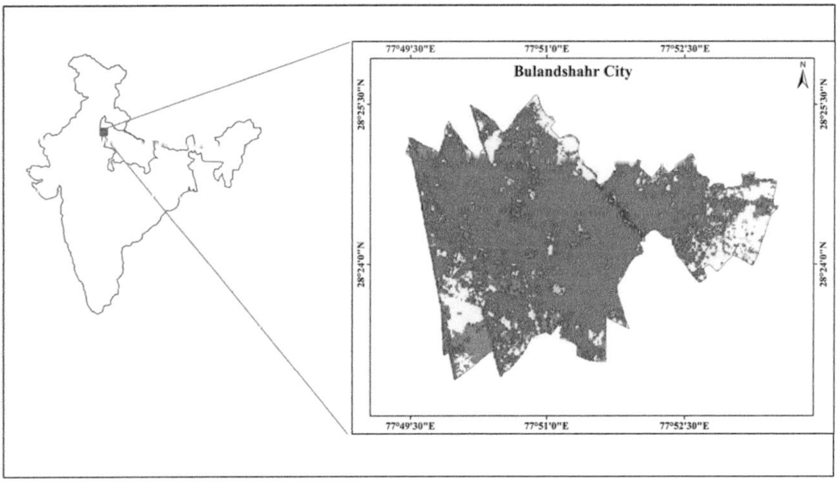

FIGURE 10.1 Study area.

10.3 SATELLITE DATA

Cloud and haze free satellite images were selected for the study. Landsat 7 ETM satellite dataset for the year 2003 (March 7th) and Landsat 8 OLI and thermal infrared sensor (TIRS) satellite dataset for the year 2018 (March 8th) have been used (Table 10.1). The satellite images have been obtained from the USGS. The preprocessing of image was carried out using ERDAS Imagine (version 14.00) software. The layer stack and extraction of area of interest has been done using survey of India toposheet of Bulandshahr city at a scale of 1:50000. All these data sets have been reprojected and brought into same projection and reformatted to 30 m cell size for analysis. In order to obtain the surface temperature data, the thermal bands of Landsat 7 ETM (2003)—specifically band 6a and Landsat 8 OLI (2018)—specifically band 10—were utilized. This was achieved by converting the digital numbers to radiance, allowing for the extraction of LST values with precision up to two decimal places. NDBI and NDVI indices have also been evaluated to analyze the relation between vegetation cover, built-up area, and LST. The distribution of land surface temperature of the city has been taken into consideration to analyze the highest and the lowest temperature of the city. Further, to show ward-based relationship, the clip tool from ARC GIS has been used to extract the values of NBVI, NDBI, and LST of different wards of the Bulandshahr city. The inclusive methodology has been shown in Figure 10.2.

TABLE 10.1 Description of Landsat Data Used in Study

Data Used	Data Acquisition Date	Number of Spectral Bands	Spatial Resolution	Source
Landsat 7 ETM/SLC	March 7th, 2003	8	30 m	https://earthexplorer.usgs.gov/
Landsat 8 OLI/TIRS	March 8th, 2018	11	30 m	

10.4 PRIMARY DATA

Field observations are made to assess the heat exposure and associated health issues such as sleep deprivation among the residents of the Bulandshahr city. All 37 administrative wards of Bulandshahr city were covered during survey. From each ward, 40 respondents were interviewed, an adult or a person above 18 years old. Total 1440 adults participated in a household survey. For better results regarding heat exposure and associated health issues, respondents were interviewed in in the year of 2018.

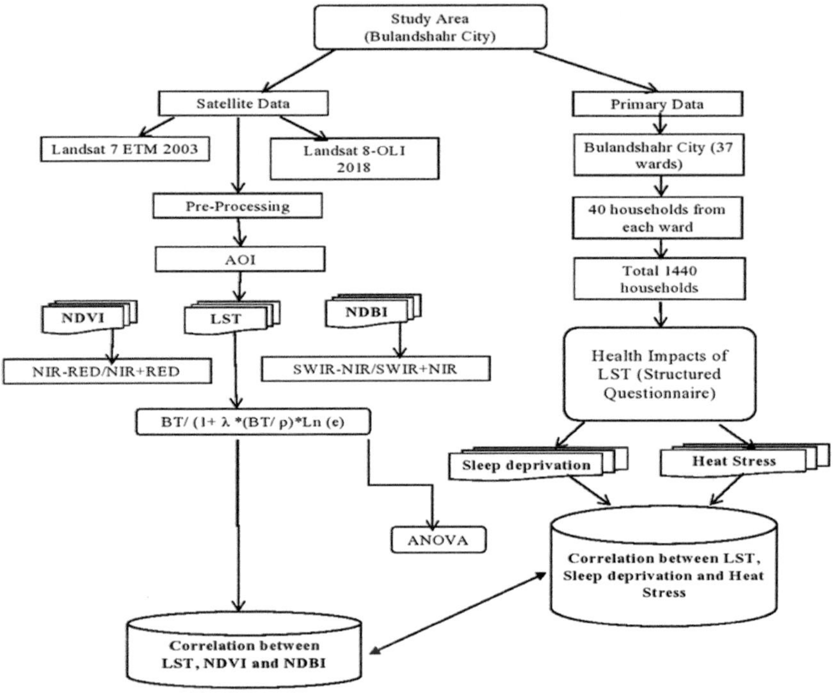

FIGURE 10.2 Study flowchart.

10.5 METHODS

10.5.1 NDBI CALCULATION

Several indexes are available to excerpt the built-up features of an area. In this analysis, NDBI is used for same. A methodology developed by Zha et al. (2003) using shortwave infrared (SWIR) and near infrared (NIR) bands is implemented to calculate NDBI (Equation 10.1). Built-up and barren lands reflect more SWIR than NIR. Water bodies do not reflect on the infrared spectrum and green surfaces reflect higher NIR than SWIR (Bahadur, 2018). The values of NDBI range from + (plus) 1 to – (minus) 1. While all +1 values present built-up areas and barren lands, values >0 show water bodies and vegetation.

$$NDBI = \frac{(SWIR - NIR)}{SWIR + NIR} \tag{10.1}$$

10.5.2 NDVI CALCULATION

NDVI is the most common index (Tucker, 1979; Purevdorj et al., 1998) used to extract vegetation proportion of the area. NDVI values were analyzed using the R and IR bands of Landsat satellites, based on the equation developed by Gao et al. (2002). NDVI has been calculated using the algorithm from Landsat 7 and 8 multitemporal images (2003 and 2018). Green area and vegetation reflect higher NIR wavelengths (Zhang et al., 2011; Liu and Weng, 2008; Mallick, 2014).

$$\mathrm{NDVI} = \frac{(\mathrm{NIR} - \mathrm{R})}{\mathrm{NIR} + \mathrm{R}} \tag{10.2}$$

10.5.3 MONO-WINDOW ALGORITHM FOR THE RETRIEVAL OF LST

Landsat 8 TIR instrument have two thermal infrared bands which are appropriate to retrieve LST with the help of split-window algorithm. Two thermal bands (bands 10 and 11) have been used in split-window algorithm obtain LST from Landsat 8 (Du et al., 2015). Several studies like Rozenstein et al. (2014) adopted the two-factor split-window algorithm for Landsat 8 data. But recently, a notice has been issued by USGS on its website regarding the uncertainty of Landsat thermal band 11 values and pointed out that users should work with TIRS band 10 data (Wang et al., 2015). In order to obtain LST, the mono-window algorithm by Qin et al. (2001), Liu and Zhang (2011), Liu and Weng (2012), and Singh et al. (2017) was applied to thermal band 10 of Landsat 8.

10.5.4 RETRIEVAL OF LST FROM LANDSAT 8

To retrieve LST, radiative transfer equation was used using band 10 of Landsat 8 TIRS data. This method was used because it has higher accuracy in comparison to single-channel method and Qin's split-window algorithm which have low and moderate accuracy, respectively (Liu and Zhang, 2011).

10.5.5 CONVERSION OF DIGITAL NUMBER TO RADIANCE

To convert digital number of band 10 into spectral radiance values, following method was used

$$\text{TOA (L)} = M_L * Q_{cal} + A_L \quad (10.3)$$

where,

M_L = band-specific multiplicative rescaling factor from the metadata (RADIANCE_MULT_BAND_*x*, where *x* is the band number).

Q_{cal} = corresponds to band 10

A_L = band-specific additive rescaling factor from the metadata (RADIANCE_ADD_BAND_*x*, where *x* is the band number).

10.5.6 THE RETRIEVAL OF BRIGHTNESS TEMPERATURE

The TIR band (Band 6) from Landsat 7 registers the radiation between the spectral range of 10.4 and 12.5 μm from the earth surface (Liu and Zhang, 2011). The thermal infrared band from Landsat 8 (Band 10) records the radiation between the spectral range of 10.6–11.9 μm from the surface of the Earth. To estimate the radiance values of thermal bands and to derive the temperature values, inverse of Plank's function was applied (Kumari et al., 2018).

$$\text{BT} = (K_2 / (\ln(K_1 / L) + 1)) - 273.15 \quad (10.4)$$

10.5.7 ESTIMATION OF NDVI

Vegetation cover of an area is an important determinant of its temperature (Weng et al., 2004); and since NDVI is most common indices to extract vegetation, in present study NDVI is calculated by using Equation 10.2 to determine the vegetation proportion of an area.

10.5.7.1 CALCULATION OF THE P_V (PROPORTION OF VEGETATION)

To calculate the P_V Landsat 8 dataset has been used. The pixel scale of vegetation cover fraction of satellite can be calculated from NDVI (Sobrino et al., 2004).

$$P_V = \text{Square}\left(\frac{(\text{NDVI} - \text{NDVI}_{min})}{(\text{NDVI}_{max} - \text{NDVI}_{min})}\right) \tag{10.5}$$

10.5.8 THE CALCULATION OF LAND SURFACE EMISSIVITY

In order to estimate the LST of an area, it is vital to estimate the land surface emissivity (ε). A Thermophysical characteristic of an object determines its emissivity. To calculate the emissivity of diverse land surface within the spectral range of 10–12 μm, the NDVI threshold technique can be used because emissivity is a variable with wavelength. Emissivity can be computed from Equation 10.6.

$$\varepsilon = 0.004 * P_v + 0.986 \tag{10.6}$$

10.5.9 RETRIEVAL OF LST

Finally, subsequent equation is utilized to convert LST from brightness temperature

$$\text{LST} = \text{BT}/\ (1 + \lambda\ *(\text{BT}/\rho)*\text{Ln}\ (e) \tag{10.7}$$

where,

BT = brightness (at satellite temperature)
λ = wavelength of emitted radiance (0.00115)
ρ = 14380 (Constant)
e = emissivity

10.5.10 RETRIEVAL OF LST FROM LANDSAT 7

The details of LST retrieval from Landsat are given in Landsat 7 handbook. The description of Landsat 7 handbook is utilized in this work to derive LST from Landsat 7 data. Following steps are used to calculate LST from Landsat 7 data:

Conversion of Digital Number into Spectral Radiance

$$L_\lambda = \text{min} + (\text{max} - \text{min})\ *\text{DN}/255\ \text{(Kumari et al., 2018)} \tag{10.8}$$

Conversion of Spectral Radiance into Sensor Radiance

$$L_\lambda = \left(\frac{L_{\mathrm{Max}(\lambda)} - L_{\mathrm{Min}(\lambda)}}{\mathrm{QCal}_{\mathrm{Max}} - \mathrm{QCal}_{\mathrm{min}}} \right) * \left(\mathrm{QCal} - \mathrm{QCal}_{\mathrm{min}} \right) + L_{\mathrm{min}\lambda} \tag{10.9}$$

(Kumari et al., 2018)
Conversion of Sensor Brightness into Sensor Radiance
The value of $K1$ and $K2$ constant is given in Table 10.2.

TABLE 10.2 Radiance Constant Value of Satellite Data (Landsat 7 and 8)

Satellite Sensor	K1	K2
Landsat 7 ETM/TRS	666.6	1260.56
Landsat 8 OLI/TIRS	774.89	1321.08

$$T = (K_2 / (\ln (K_1 / L) + 1)) \tag{10.10}$$

In the final step of LST calculation for obtaining results in Celsius, the temperature is changed from Kelvin to degree centigrade by Equation 10.11 using the following formula:

$$T\,(^\circ\mathrm{C}) = T\,(\mathrm{K}) - 273.15 \tag{10.11}$$

10.5.11 KARL PEARSON'S CORRELATION COEFFICIENT

Correlation is used to analyze the relationship of LST with NDVI and NDBI and with health indicators like sleep deprivation and heat stress.

$$r = \frac{n\Sigma x_i y_i - \Sigma x_i y_i}{\sqrt{n\Sigma x_i^2 - (\Sigma x_i)^2} \times \sqrt{n\Sigma y_i^2 - (\Sigma y_i)^2}} \tag{10.12}$$

10.6 RESULTS

10.6.1 VEGETATION PATTERN OF BULANDSHAHR CITY

Urbanization is a global phenomenon, characterized by the conversion of agricultural land into developments and the gradual disappearance of natural vegetation in cities due to new construction projects. Landsat 7 and 8 imageries for the year 2003 and 2018 were extracted to analyze and generalize + the change matrix of NDVI for study area (Figures 10.3 and 10.4). The NDVI images were further reclassified into five classes, representing density of vegetation cover in the city, ranging from very low (<0.1), low (0.1–0.2), medium (0.2–0.3), high (0.3–0.4), and very high (>0.4) (Tables 10.3 and 10.4). The results concluded that the year 2003 has higher values of NDVI in comparison to the year 2018. The higher values of NDVI were observed around the N, NE, SW, and W part of the city, and the lowest values of NDVI were observed in the central, built-up dominant residential area of the city. In 2003, the NDVI values spanned from −0.29 to 0.59. However, in 2018, the NDVI values decreased, ranging from −0.26 to 0.49. The vegetation pattern of the city suggests that there is a dominance of vegetation and agriculture area under the high NDVI value of the city. The area under the lowest NDVI values is the area where high built-up area is dominating.

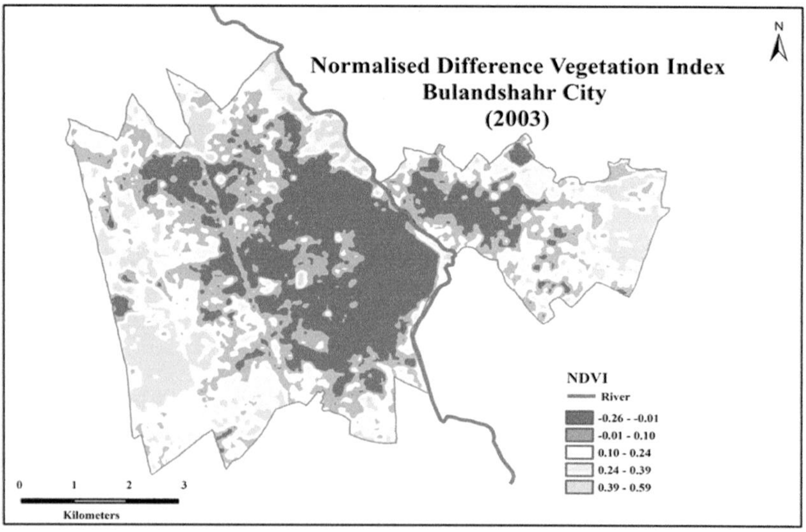

FIGURE 10.3 Landsat-derived NDVI image of study area (2003).
Source: Landsat 7 ETM/SLC (30 m Resolution).

While comparing NDVI values from 2003 and 2018 images, most significant changes are observed in the low and very high categories. Area under very high values of NDVI has been decreased by 70.87% in the year

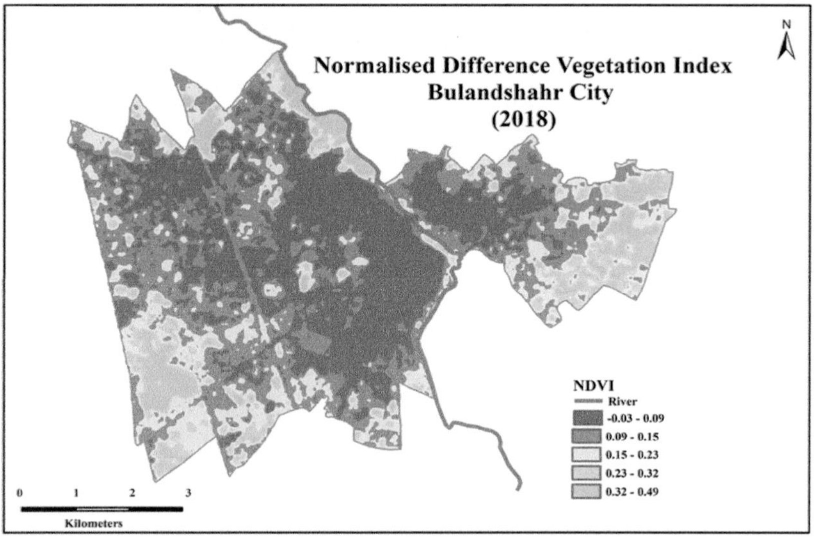

FIGURE 10.4 Landsat-derived NDVI image of study area (2018).

Source: Landsat 8 OLI/TIRS (15-m resolution).

TABLE 10.3 Reclassified Values of NDVI From 2003 to 2018

NDVI Density Classes		2003		2018		Change Between 2003 and 2018
		Sq. km	Percentage	Sq. km	Percentage	
Very low	<0.1	8.68	52.08	6.92	41.49	−11.32
Low	0.1–0.2	2.71	16.28	6.56	39.36	41.48
Medium	0.2–0.3	2.04	12.26	1.69	10.16	−9.39
High	0.3–0.4	1.43	8.58	1.2	7.17	−8.95
Very high	>0.4	1.8	10.8	0.31	1.84	−70.88
Total		16.67	100	16.68	100	

TABLE 10.4 Change in NDVI Area from 2003 to 2018

Category	Area Percentage		Changes Between 2003 and 2018
	2003	2018	
Very low	7.58	5.87	−1.71
Low	23.83	8.19	−15.64
Medium	25.63	13.13	−12.49
High	32.28	37	4.71
Very high	10.66	35.79	25.12
Total	100	100	0

2018, while area under low NDVI density has been increased by 41.47% in the same year suggesting a clear loss of vegetation in the study area within this period. The decrease of land in the very high category of NDVI values is the result of increase in built-up area, encroached vegetation, and agricultural lands of the city. The NDVI analysis has made it clear that proportion of vegetation in the city is shrinking, leading to decline in the evaporative cooling and adding to rising surface temperature.

10.6.2 PATTERN OF BUILT-UP AREA

Another most important variable which is affecting land surface temperature of the Bulandshahr city is augmentation of built-up surfaces in the city; therefore, to document any change in built-up area in the city, NDBI has been computed for the years 2003 and 2018 (Figures 10.5 and 10.6). The results revealed that positive changes were observed between the years of 2003 and 2018 in built-up density. The NDBI values in 2003 were recorded from −0.46 to 0.22 which in 2018 have increased from −0.57 to 0.36. It is indicating a remarkable upsurge in built-up area between the years 2003 and 2018.

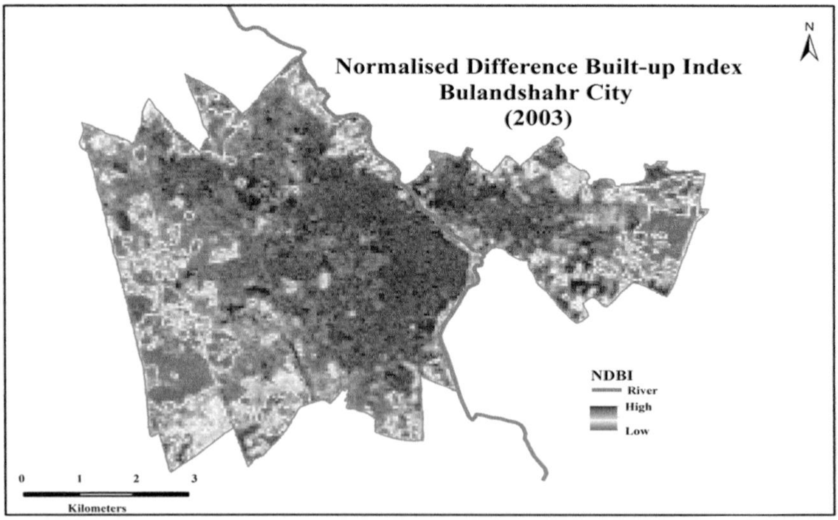

FIGURE 10.5 Landsat-derived NDBI image of study area (2003).

Source: Landsat 7 ETM/SLC (30-m resolution).

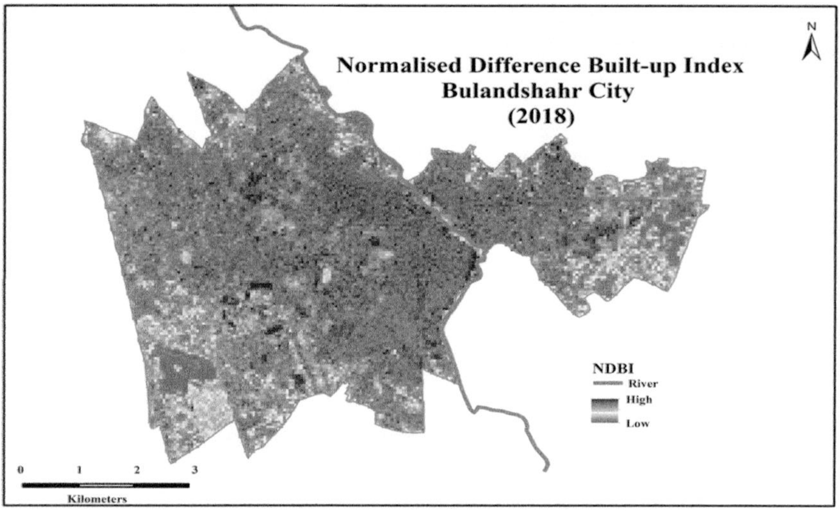

FIGURE 10.6 Landsat-derived NDBI image of study area (2018).
Source: Landsat 8 OLI/TIRS (15-m resolution).

In the year 2003, higher values of NDBI were observed in the middle part of the city which is also the oldest part of the city where low income residential population resides. Furthermore, CBD of the Bulandshahr city is also located in this part where availability of urban services and infrastructural facilities is higher than the other part of the city. However, for the year 2018, higher values of NDBI are well distributed throughout the city except extreme corners of the city which is east, north, and south-westers parts. It indicates that built-up area has increased drastically from 2003 to 2018. It is also important to note that vacant lands of the city have recorded higher NDBI values due to their higher reflectance in the SWIR region. Since vacant lands in the city are encroachment upon agricultural and green lands for the construction of apartments and other residential units, they have also contributed in increasing built-up region of the city.

10.6.3 THE SPATIOTEMPORAL PATTERN OF LST

The LST values of Bulandshahr city for the year 2003 and 2018 were quantitatively retrieved from ETM/OLI (TIRS) images to evaluate spatio-temporal patterns of the study area (Figures 10.7 and 10.8). The LST values found in Bulandshahr city ranges from 26.61 to 35.96 for the year 2003, and 26.44 to 37.07 for the year 2018. These ranges were further grouped in very low to very high categories (Table 10.5). In the year 2003, total area under very low LST category was 8%, whereas in 2018, total area under very low LST category was observed almost 6%. In low LST category, total area has decreased by −16%, whereas in very high LST category total area has increased by 25% between the years 2003 and 2018. In 2003, it was observed that small central part of the city has higher land surface value but in 2018 the LST values have increased drastically and covered almost entire city. The core of city is prone to increased temperature due to urbanization which had taken place in Bulandshahr city over the period of 16 years (Figures 10.7 and 10.8). The highest LST value was observed in 2018 in comparison to 2003 and occupied almost southeastern and central part of the city, whereas in the year 2003, small patches of high temperature appeared in the north and northwestern part of the city. Based on the cross-section profile of the city illustrated in Figure 10.9, it was noted that the disparity in surface radiance temperature between developed areas and vegetated areas can be as high as 7°C.

TABLE 10.5 Change in LST Values from 2003 to 2018

Category	Area Percentage		Changes Between 2003 and 2018
	2003	2018	
Very low	7.58	5.87	−1.71
Low	23.83	8.19	−15.64
Medium	25.63	13.13	−12.49
High	32.28	37	4.71
Very high	10.66	35.79	25.12
Total	100	100	0

The temporal variation in surface temperature of the city has been shown in Figure 10.10, representing the mean values of LST in both the years.

10.6.4 RELATIONSHIP BETWEEN NDVI, NDBI, AND LST

A ward-wise spatiotemporal analysis of NDVI, LST, and NDBI was done to analyze the impact of built-up and vegetation areas on the LST. To calculate the association between the factors, mean values of NDVI,

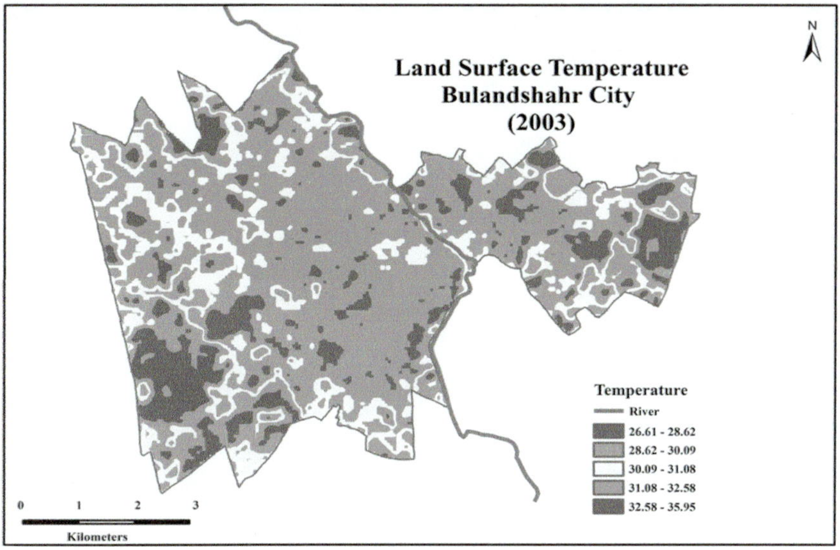

FIGURE 10.7 Landsat-derived LST image of study area (2003).

Source: Landsat 7 ETM/SLC (30-m resolution).

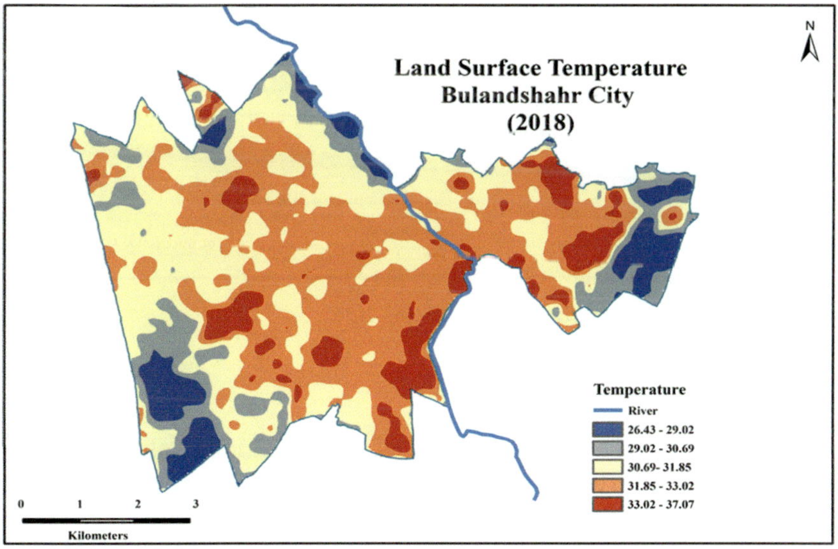

FIGURE 10.8 Landsat-derived LST image of study area (2018).

Source: Landsat 8 OLI/TIRS (15-m resolution).

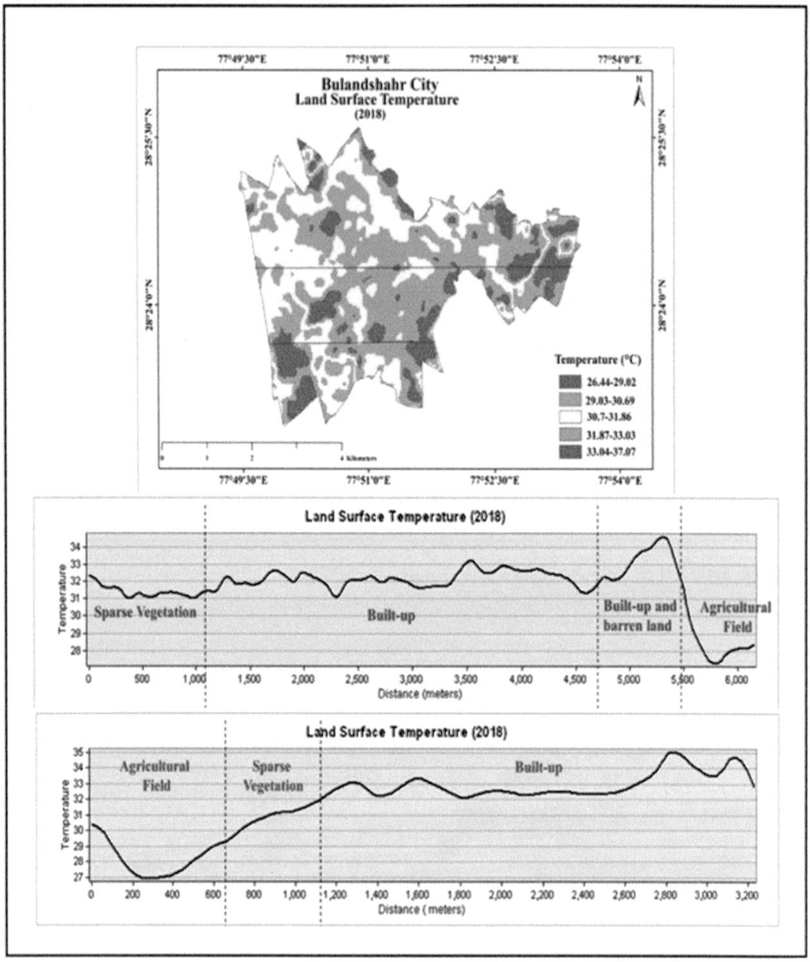

FIGURE 10.9 Spatial profile of LST 2003 and 2018.

Source: Locations of transect line profile.

NDBI, and LST of each ward was used. The maximum values of NDVI were found in the western part of the city in ward 10, that is, 0.32 for the year 2003 (Table 10.6). In 2003, agricultural/vegetation cover accounted for 46.57% of the city's total area. By 2018, this percentage had decreased to 27.74% (Table 10.6). The study shows that the areas which recorded high NDVI values have shown a slightly less land surface temperature in comparison to areas of low NDVI values.

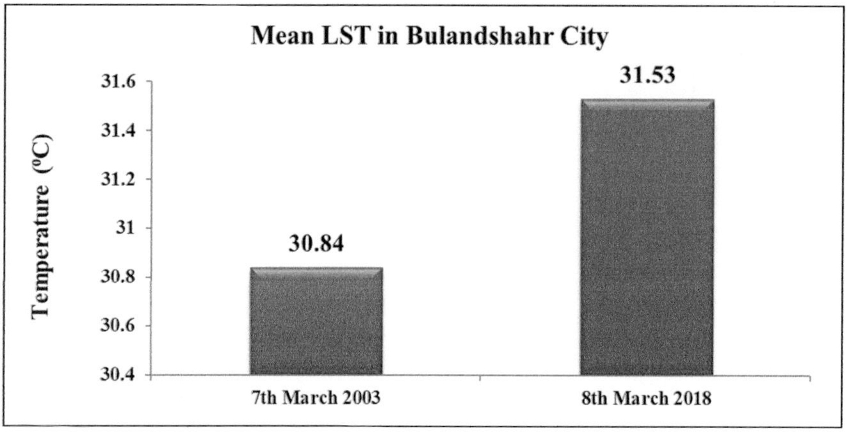

FIGURE 10.10 Mean temperature of study area (2003 and 2018).

NDBI was used to calculate the built-up density of each ward of the city. The central parts of the city have shown high built-up density. Ward no. 20 of Bulandshahr city has recorded highest mean value, which is 0.07, while ward no. 10 recorded lowest NDBI value, that is, −0.21, it is the same ward where highest NDVI values are recorded for year 2003 (Table 10.7). LST values are also calculated for each ward to analyze the association between variables. The lowest values for LST were recorded in ward number 10 in which lowest values of NDVI were also recorded, and it is evidently illustrating the whole scenario of relationships between NDVI, NDBI, and LST.

Year 2018 has witnessed a drastic change in land use/land cover pattern in the city because of the various ongoing development projects. The results indicate a scenario of total deterioration in the green areas of the city. The ward no. 17 had the highest mean NDVI values in 2018 at 0.26, whereas the wards no. 14 and 26, which include the CBD of the city, recorded the lowest mean NDVI values. From Table 10.8, it can be concluded that the mean values of NDBI of the city were recorded in the ward no. 4 which is the densely populated area of the city. The significant change in vegetation and built-up area between the years of 2003 and 2018 demonstrates the rapid urban growth in the study area. The study reveals that the LST mean values were found higher in the low NDVI areas, which is responsible for major changes in local climate of the city.

TABLE 10.6 Mean Values of LST, NDVI, and NDBI of Each Ward for Year 2003

2003												
Wards	**LST**				**NDVI**				**NDBI**			
	Mean	**Max**	**Min**	**Sd**	**Min**	**Max**	**Mean**	**SD**	**Min**	**Max**	**Mean**	**SD**
1	30.92	33.05	27.12	1.22	−0.11	0.55	0.04	0.17	−0.44	0.11	−0.01	0.13
2	30.47	34.01	26.61	1.55	−0.09	0.58	0.17	0.18	−0.46	0.16	−0.08	0.15
3	30.58	32.59	27.63	1	−0.08	0.54	0.17	0.13	−0.4	0.13	−0.07	0.11
4	31.83	32.59	30.63	0.5	−0.11	0.04	−0.08	0.01	0.03	0.1	0.06	0.01
5	31.17	33.06	29.14	0.66	−0.1	0.51	0.03	0.1	−0.4	0.13	0	0.07
6	31.26	35.96	28.63	1.37	−0.09	0.48	0.14	0.15	−0.34	0.11	−0.07	0.11
7	31.85	32.59	31.12	0.4	−0.12	0	−0.09	0.02	0	0.1	0.06	0.01
8	30.26	33.05	27.63	1.2	−0.06	0.54	0.2	0.13	−0.42	0.11	−0.09	0.12
9	31.8	35.48	29.14	0.95	−0.1	0.45	0.08	0.12	−0.33	0.13	0	0.08
10	28.97	32.1	26.61	1.2	−0.05	0.58	0.32	0.15	−0.43	0.11	−0.21	0.12
11	31.75	33.07	21.12	0.46	−0.12	0.16	−0.05	0.05	−0.09	0.12	0.05	0.02
12	31.74	33.06	30.13	0.59	−0.1	0.44	0	0.11	−0.25	0.11	0.02	0.07
13	32.07	33.07	31.12	0.42	−0.11	0.06	−0.06	0.03	−0.18	0.14	0.05	0.03
14	31.19	31.61	30.63	0.32	−0.14	−0.03	−0.11	0.01	0.03	0.09	0.06	0.01
15	31.4	33.56	28.63	0.7	−0.13	0.46	0.03	0.12	−0.28	0.12	0.01	0.08
16	31.46	33.55	29.14	0.68	−0.12	0.5	0.01	0.08	−0.33	0.21	0.02	0.06
17	29.82	35.01	26.61	1.61	−0.05	0.57	0.28	0.17	−0.44	0.12	−0.16	0.15
18	31.8	33.06	29.63	0.52	−0.12	0.06	−0.08	0.04	−0.03	0.11	0.06	0.02
19	30.93	33.06	28.13	0.92	−0.13	0.48	0.11	0.13	−0.37	0.12	−0.04	0.1

TABLE 10.6 *(Continued)*

	2003											
Wards	LST				NDVI				NDBI			
	Mean	Max	Min	Sd	Min	Max	Mean	SD	Min	Max	Mean	SD
20	31.85	32.59	31.12	0.48	−0.11	−0.09	−0.08	0.01	0.03	0.13	0.07	0.01
21	30.66	33.56	27.12	1.41	−0.09	0.59	0.24	0.16	−0.44	0.12	−0.11	0.14
22	32.05	33.07	31.12	0.44	−0.13	0.06	−0.08	0.03	−0.02	0.12	0.05	0.02
23	31.62	33.06	29.14	0.62	−0.12	0.32	−0.02	0.07	−0.26	0.16	0.03	0.05
24	30.6	35.96	27.12	1.76	−0.26	0.55	0.18	0.15	−0.43	0.14	−0.1	0.12
25	31.72	33.06	29.63	0.62	−0.14	0.47	0	0.09	−0.24	0.1	0.03	0.05
26	31.56	32.58	30.62	0.31	−0.14	−0.06	−0.1	0.01	0.02	0.13	0.06	0.01
27	31.96	34.53	29.63	0.85	−0.12	0.36	0.01	0.09	−0.29	0.11	0.01	0.06
28	30.58	32.59	27.63	1.2	−0.12	0.51	0.11	0.21	−0.38	0.11	−0.06	0.15
29	30.4	33.05	27.63	0.95	−0.09	0.53	0.13	0.12	−0.41	0.11	−0.06	0.1
30	31.75	33.56	31.12	0.44	−0.14	0.08	−0.09	0.03	0	0.12	0.07	0.02
31	31.56	33.06	30.13	0.67	−0.12	0.4	0.03	0.09	−0.23	0.12	0	0.06
32	31.95	34.53	29.14	0.85	−0.12	0.51	0.03	0.11	−0.42	0.13	0.02	0.07
33	31.74	32.59	30.63	0.4	−0.13	0.04	−0.09	0.02	−0.01	0.11	0.06	0.02
34	31.42	34.02	27.63	1.38	−0.16	0.41	0.05	0.12	−0.31	0.18	0	0.1
35	31.64	33.56	29.63	1.09	−0.09	0.48	0.11	0.16	−0.32	0.09	−0.03	0.12
36	31.46	33.06	29.63	0.69	−0.12	0.39	0.01	0.14	−0.27	0.13	0.02	0.09
37	30.96	32.11	30.13	0.58	−0.14	−0.04	−0.1	0.02	0.02	0.11	0.06	0.01

TABLE 10.7 Mean Values of LST, NDVI, and NDBI of Each Ward for Year 2018

Wards	2018											
	LST				NDVI				NDBI			
	Mean	Max	Min	Sd	Mean	Max	Min	Sd	Mean	Max	Min	Sd
1	31.65	33.18	28.85	0.78	0.09	0.42	0.04	0.07	−0.02	0.04	−0.28	0.05
2	31.44	34.35	27.54	1.53	0.17	0.49	0.05	0.11	−0.06	0.07	−0.35	0.1
3	31.33	33.76	27.54	1.53	0.12	0.33	0.04	0.05	−0.03	0.05	−0.23	0.03
4	31.47	32.09	31.08	0.31	0.05	0.1	0.03	0.01	0.01	0.03	0.01	0.01
5	31.99	33.13	31	0.48	0.08	0.2	0.02	0.03	−0.01	0.04	−0.07	0.02
6	32.4	34.27	28.24	0.99	0.13	0.38	0.04	0.05	−0.03	0.05	−0.26	0.05
7	32.67	33.74	31.95	0.37	0.05	0.09	0.03	0.01	0.01	0.04	−0.01	0.01
8	31.24	32.25	29.07	0.67	0.15	0.44	0.05	0.07	−0.04	0.07	−0.32	0.07
9	32.48	36.56	30.12	0.94	0.13	0.4	0.04	0.06	−0.02	0.06	−0.25	0.05
10	29.92	32.74	26.44	1.57	0.22	0.46	0.04	0.1	−0.11	0.06	−0.35	0.1
11	32.38	34.77	31.7	0.48	0.06	0.24	0.03	0.03	0	0.05	−0.12	0.01
12	31.86	33.11	30.1	0.72	0.08	0.16	0.04	0.03	0.01	0.06	−0.04	0.02
13	32.84	34.93	32.06	0.64	0.06	0.14	0.03	0.02	0	0.04	−0.03	0.01
14	31.79	32.28	31.55	0.18	0.04	0.11	0.03	0.01	0.02	0.04	−0.02	0.01
15	31.61	32.4	30.01	0.47	0.11	0.44	0.03	0.07	−0.02	0.04	−0.3	0.05
16	32.24	34.04	31.15	0.53	0.11	0.31	0.03	0.04	−0.01	0.07	−0.14	0.03
17	29.96	37.05	26.69	1.84	0.26	0.46	0.07	0.09	−0.13	0.07	−0.34	0.11
18	32.17	32.42	31.45	0.17	0.05	0.13	0.03	0.02	0.02	0.06	−0.02	0.01
19	32.21	34.99	29.83	0.9	0.13	0.38	0.04	0.06	−0.02	0.07	−0.26	0.05

TABLE 10.7 *(Continued)*

Wards	2018											
	LST				NDVI				NDBI			
	Mean	Max	Min	Sd	Mean	Max	Min	Sd	Mean	Max	Min	Sd
20	32.56	33.21	32.06	0.27	0.05	0.1	0.04	0.01	0.01	0.03	−0.02	0.01
21	30.17	32.64	27.06	1.41	0.22	0.48	0.05	0.13	−0.1	0.05	−0.37	0.11
22	32.26	33.24	31.82	0.31	0.06	0.14	0.03	0.27	0	0.05	−0.05	0.02
23	32.1	33.11	31.03	0.41	0.09	0.26	0.03	0.04	0	0.06	−0.01	0.03
24	31.72	35.11	29.06	1.23	0.15	0.43	−0.04	0.07	−0.04	0.11	−0.32	0.06
25	32.14	33.69	30.91	0.46	0.11	0.42	0.02	0.07	−0.01	0.09	−0.25	0.04
26	31.78	32.25	31.13	0.21	0.04	0.06	0.03	0.01	0.02	0.04	0	0.01
27	32.64	35.96	30.69	0.99	0.1	0.3	0.03	0.05	0	0.09	−0.18	0.04
28	30.15	31.85	27.57	1.3	0.17	0.46	0.03	0.13	−0.72	0.04	−0.33	0.11
29	31.57	32.96	30.64	0.45	0.1	0.3	0.32	0.04	−0.01	0.06	−0.16	0.03
30	32.38	33.25	31.86	0.24	0.05	0.13	0.03	0.02	0	0.04	−0.06	0.01
31	32.43	33.84	31.06	0.44	0.09	0.23	0.03	0.04	−0.02	0.05	−0.1	0.02
32	31.67	33.9	29.78	0.67	0.1	0.31	0.04	0.05	−0.01	0.08	−0.02	0.03
33	32.16	32.95	31.15	0.28	0.05	0.15	0.03	0.02	0.01	0.04	−0.05	0.02
34	33.14	34.74	32.16	0.57	0.11	0.38	0.03	0.07	0	0.07	−0.22	0.04
35	33.41	34.28	32.16	0.57	0.13	0.28	0.03	0.08	0	0.08	−0.13	0.05
36	32.37	33.25	31.6	0.43	0.1	0.35	0.03	0.08	0	0.06	−0.15	0.04
37	31.72	32.47	31.38	0.31	0.14	0.49	0.04	0.09	0.01	0.04	−0.02	0.01

TABLE 10.8 Correlations Between LST, NDVI, and NDBI Between 2003 and 2018

	2003				2018		
LST	**LST**	**NDVI**	**NDBI**	**LST**	**LST**	**NDVI**	**NDBI**
	1				**1**		
NDVI	−0.815**	1		NDVI	−0.668**	1	
NDBI	0.881**		1	NDBI	0.550**	−0.446**	1

The association among LST, NDVI, and NDBI were established for years 2003 and 2018 (Table 10.8). From the analysis, it was found that the areas which have recorded highest values of NDBI also recorded the high surface temperatures and the areas lying under high NDVI values have lower surface temperatures in comparison to areas of low NDVI values.

10.6.5 LST AND HEALTH

Increased temperature due to climate change is one of the crucial factors that affect human health and their quality of life (Tan et al., 2010; Jenerette et al., 2016; Wong et al., 2017). However, in the present study, an attempt was made to prove this hypothesis whether increased temperature affect human health or not. Therefore, with the help of the statistical analysis, the above hypothesis was tested. The fundamental difference between present and previous studies is the use of a questionnaire interview along with satellite data. The questionnaire surveys can provide more accurate picture of the reality, therefore, household survey of Bulandshahr city was done. In this study, primary data for sleep deprivation and heat stress has been collected through questionnaire survey in the year 2018. To inspect the impact of LST on heat stress in Bulandshahr city, Karl Pearson's Correlation has been adopted. Data were analyzed and significance of their correlation was tested. It became clear from the analysis that increasing surface temperature is causing sleeping problems which is significant at 1%. A strong positive correlation was observed between the occurrence of sleep deprivation (r=0.480**) and increased land surface temperature. There is a positive correlation between heat stress and increasing land surface temperature, but the relationship is not deemed statistically significant (r= 0.216) (Table 10.9). The evaluation of the spatial dynamics of the self-reported health in the Bulandshahr city was done through the process of mapping, therefore, GIS was used to dictate this dynamic.

TABLE 10.9 Correlation Between LST and Health (Sleep Deprivation and Heat Stress) (2018)

Associated Problem	Land Surface Temperature
Sleep deprivation	0.480**
Heat stress	0.216

10.6.5.1 LST AND SLEEP DEPRIVATION

The spatial distribution of sleep deprivation data of Bulandshahr city indicates that the areas of high-temperature distribution have a close link with sleep deprivation data. The older or central part of the city has high population density as well as high temperature which indicate that increased temperature can affect the sleep quality of the residents. Civil lines area which is the high-population density area of the city, sleep quality distribution was observed in low percentage, indicating that the better infrastructural facilities in an area can also affect the quality of life of the residents. These are the areas of high income population which indicates the better living conditions can also affect the mental health responses of a residential area (Figure 10.11).

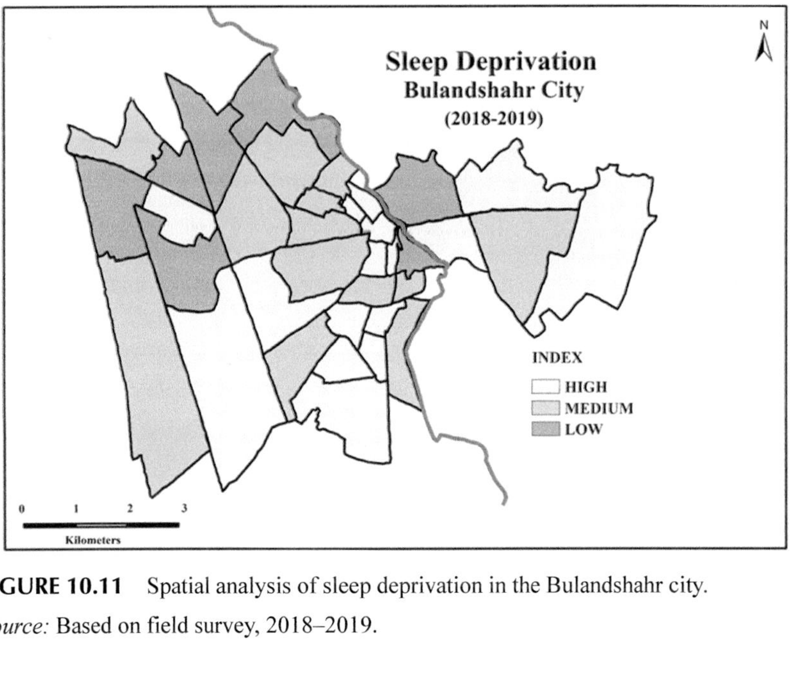

FIGURE 10.11 Spatial analysis of sleep deprivation in the Bulandshahr city.

Source: Based on field survey, 2018–2019.

10.6.5.2 LST AND HEAT STRESS

Increased nighttime temperature that is higher than normal body temperature, that is, 37°C can trigger psychological stress in human body. If these circumstances continue for a long period of time more, it may affect the vulnerable population of the city (Taha, 1996; Rosenfeld et al., 1998; Akbari et al., 1997). The analysis of the spatial distribution of the heat stress in Bulandshahr city highlights the same dynamics as sleep deprivation (Figure 10.12). Medium and high distribution areas have shown a close link to high temperature algorithms. Meanwhile, it can be argued that not only physical attributes, but also improved financial opportunities can impact the outcome dynamics.

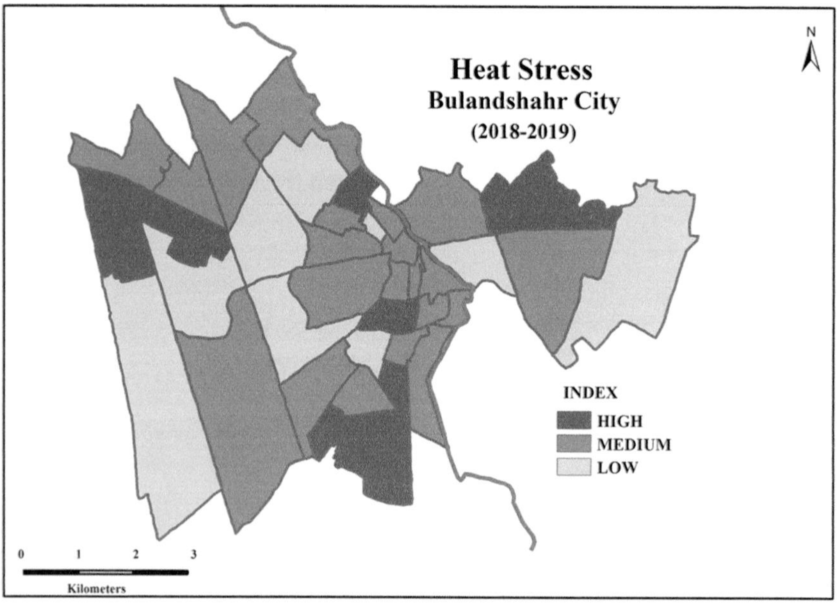

FIGURE 10.12 Spatial analysis of population density in Bulandshahr city (2018).

10.7 DISCUSSION

To assess spatiotemporal distribution of urban thermal environment, LST is considered one of the most effective methods. It is evident that the change in urban microclimate is the result of increasing manmade environment and transformation of natural vegetation. The calculated land surface temperature of the Bulandshahr ranged between 26.61°C and 35.95°C in 2003 with the mean value of 30.82°C, and in year 2018, it ranged between 26.44°C and 37.08°C with mean value of 31.52°C. The maximum temperature was recorded in the older/mid part because land use pattern of older part of the Bulandshahr city mainly consists of the asphalt paved roads and other anthropogenic conversion of land into Industrial, commercial residential and transportation area. As per Chen et al. (2006), change in LST of any city is mainly influenced by natural landscape, that is, land use, topography; distribution of green cover, built-up, open land, water bodies. Present study demonstrated a correlation between LST, NDVI, and NDBI using Karl Pearson's correlation technique using SPSS software. It was observed that in the year 2003, there was positive correlation between NDBI and LST, whereas between LST and NDVI, negative correlation was observed. In 2003, a high degree correlation (r^2=0.881) between LST and NDBI has been observed and also a high degree negative correlation (r^2=−0.815) observed between LST and NDVI. In comparison to 2003, the correlation among LST, NDBI, and NDVI in 2018 was of weak degree, that is, the relation between LST and NDBI was observed of medium degree (r^2=0.550), whereas relation between LST and NDVI (r^2=−0.668) was observed negatively correlated. A positive correlation between LST and NDBI indicates that there is a continuous increase in the temperature in the areas of high population density. The result of the ANOVA test has also suggested that estimated LST has significantly increased over the years.

In the present work, LST was used for quantifiable depiction of heat island effect on surface temperature of the city. The spatiotemporal evaluation of NDBI and NDVI clearly indicates that urbanization and degradation of green environment are the major causes of change in LST value of the Bulandshahr city and an increasing trend of minimum/maximum temperature. Furthermore, it is evident from several studies which were executed by the researchers worldwide, already discussed in literature review/introduction of the study. Several research reports have indicated that the rise in urban temperatures and alteration in both local and microclimates are predominantly attributed to the haphazard implementation of development projects in urban areas. Another important factor behind the elevated LST is the conversation of natural vegetation into concrete jungle. So, anthropogenic activities such as clearing of vegetated cover and conversion of open space into built-up area are playing an important part. Finally, study concludes that urbanization and increasing built-up areas have direct effect on the microclimate conversion in the study area.

To examine the impact of LST on health (sleep deprivation and heat stress) of the people in the present study area Karl Pearson's correlation was adopted for analysis and significance of their correlation is tested. A strong positive correlation was observed between the occurrence of sleep issues (r=0.480**) and increased land surface temperature. Although stress shows positive correlation but there is no significant relation between increasing LST and stress (r=0.216). Wards located in the high population density areas and generally related to economically weaker section showed high rates of sleep disturbance, whereas heat stress analysis showed no significant correlation with LST in the present study. Remote sensing data has been used extensively over a significant period to analyze the phenomenon of heat islands, which typically entail higher urban temperatures in contrast to the surrounding rural areas (Voogt and Oke, 2003). However, UHI or LST has never been used to study heat stress but it has a great potential to study spatial heat stress environment in urban areas (Song and Wu, 2018). Past studies elaborated heat stress in urban areas using single indices (Belding and Hatch, 1955; Dufton, 1929; Wallace et al., 2005) using RS images. However, present study has used RS data and self-reported health problems to study the level of discomfort among the residents of the Bulandshahr city. Several other studies related to heat risk suggest that the adaptation of human body to increased temperatures has an upper limit which makes it important to study the consequences of increase temperature on human body (Sherwood and Huber, 2010). Several previous studies have identified that the exposure to high thermal temperature can lead to heat rash and heat exhaustion not only this but it can also impact circulatory system and mental health as well (Cheshire, 2016; Gasparrini et al., 2015; Mullins and White, 2019). Elevated high temperature can affect the psychological responses of human body, further these psychological responses can disturb the quality of sleep, biological circadian rhythm, and cardiovascular system (Murage et al., 2017; Thosar et al., 2018).

Various other studies suggest that the population density of an area is often related to increased temperatures in urban areas and generally considered prone to health risks (Meze-Hausken, 2008). With explicit reference to heat stress, many studies argue that high population density in urban area can increase the health risk in residing population, hence, it is practical to include the people of high density area as a vulnerable group (Harlan et al., 2006; Dolney and Sheridan, 2006; Hajat and Kosatky, 2010). Moreover, the escalation of global temperatures can lead to hazardous weather conditions, resulting in an elevated risk of heat stress, as well as psychological and physical ailments (Hulme, 2014; Rifkin et al., 2018). Another limitation in coping with heat-related risk factors can be the congested or nonavailable green spaces, poorly constructed area that hinders air flow and disrupt thermal flow of an area. Financial constraints can also play an important role to deal with heat related coping strategies.

Many scholars around the globe have conducted research regarding the effect of UHI, and argue that the UHI phenomena is the result of many factors which are closely related to heat release, vegetation cover, structure of urban surface, weather conditions, and population density (Arnfield, 2003; Kotharkar and Surawar, 2016; Oke and Maxwell, 1975; Rizwan et al., 2008; Yang et al., 1975, 2016; Weng et al., 2004). Meanwhile, the spatial analysis of Bulandshahr city has revealed that the central and older parts of the city have the highest population density while areas such as Shivpuri, Chandpur, and Surya Nagar which are comparatively new wards of the city, have shown low population density. From the analysis of LST distribution, it can be observed that in the year 2003, the high-temperature area was noted in the central part of the city, later, it spread across the city border. Furthermore, a close analysis between heat stress, sleep deprivation, and population density has been done. It clearly indicates that the population density of an area can affect the heat sensitivity among the residents of the particular area.

10.8 CONCLUSION

Present study provides an overview of microclimatic conditions of the Bulandshahr city and its impact on human health. Indices such as NDVI, NDBI, and LST were estimated of study area using satellite images of Landsat 7 and 8, further, their relationships with each other were examined. In the same study, an attempt has been made to study the impact of UHI on human health. Therefore, a relationship between LST and self-reported health problem were analyzed. The findings indicate that the expansion of urban areas and the transformation of green spaces into residential and commercial development have resulted in a rise in the city's land surface temperature. Further investigation proved that the vegetation covers showed positive correlation in regulating temperature of study area. Self-reported psychological responses related to increased temperature were also observed and it was found that residents related to economically weaker section of society are more vulnerable to heat-related issues. However, one of the major limitations or disadvantages of the present study is spectral resolution of the data as it is difficult to obtain the data at microscale. Fine resolution data with ground centered thermal sensors are also limits of this work. Nevertheless, the research offers insights into how urban development affects LST in urban areas, and how it can impact human health.

KEYWORDS

- **land surface temperature**
- **normalized difference vegetation index**
- **remote sensing and GIS**
- **urban microclimate**
- **sleep deprivation**
- **heat stress**

REFERENCES

Akbari, H., Bretz, S., Kurn, D. M., and Hanford, J. (1997). Peak power and cooling energy savings of high-albedo roofs. *Energy and Buildings 25*(2), 117–126. https://doi.org/10.1016/s0378-7788(96)01001-8

Anderson, G. B., and Bell, M. L. (2011). Heat waves in the United States: Mortality risk during heat waves and effect modification by heat wave characteristics in 43 US communities. *Environmental Health Perspectives 119*(2), 210–218.

Arnfield, A. J. (2003). Two decades of urban climate research: A review of turbulence, exchanges of energy and water, and the urban heat island. *International Journal of Climatology: A Journal of the Royal Meteorological Society 23*(1), 1–26.

Bahadur. (2018). NDVI, NDBI & NDWI Calculation Using Landsat 7,8. Https://Www.Linkedin.Com/. Retrieved from https://www.linkedin.com/pulse/ndvi-ndbi-ndwi-calculation-using-landsat-7-8-tek-bahadur-kshetri/%0Ahttps://www.linkedin.com/pulse/ndvi-ndbi-ndwi-calculation-using-landsat-7-8-tek-bahadur-kshetri (Accessed (September 8, 2018).

Bambrick, H. J., Capon, A. G., Barnett, G. B., Beaty, R. M., and Burton, A. J. (2011). Climate change and health in the urban environment: Adaptation opportunities in Australian cities. *Asia Pacific Journal of Public Health 23*(2 Suppl), 67S–79S.

Belding, H. S., and Hatch, T. F. (1955). Index for evaluating heat stress in terms of resulting physiological strains. *Heating, Piping and Air Conditioning 27*(8), 129–36.

Chen, X. L., Zhao, H. M., Li, P. X., and Yin, Z. Y. (2006). Remote sensing image-based analysis of the relationship between urban heat island and land use/cover changes. *Remote Sensing of Environment 104*(2), 133–146. https://doi.org/10.1016/j.rse.2005.11.016

Cheshire Jr, W. P. (2016). Thermoregulatory disorders and illness related to heat and cold stress. *Autonomic Neuroscience 196*, 91–104.

Dolney, T. J., and Sheridan, S. C. (2006). The relationship between extreme heat and ambulance response calls for the city of Toronto, Ontario, Canada. *Environmental Research 101*(1), 94–103. https://doi.org/10.1016/j.envres.2005.08.008

Du, C., Ren, H., Qin, Q., Meng, J., and Zhao, S. (2015). A practical split-window algorithm for estimating land surface temperature from landsat 8 data. *Remote Sensing 7*(1), 647–665. https://doi.org/10.3390/rs70100647

Dufton, A. F. (1929). The eupatheostat. *Journal of Scientific Instruments 6*(8), 249.

Gao, F., Jin, Y., Li, X., Schaaf, C. B., and Strahler, A. H. (2002). Bidirectional NDVI and atmospherically resistant BRDF inversion for vegetation canopy. *IEEE Transactions on Geoscience and Remote Sensing 40*(6), 1269–1278. https://doi.org/10.1109/TGRS.2002.800241

Gasparrini, A., Guo, Y., Hashizume, M., Lavigne, E., Zanobetti, A., Schwartz, J., ... and Armstrong, B. (2015). Mortality risk attributable to high and low ambient temperature: A multicountry observational study. *The Lancet 386*(9991), 369–375.

Hajat, S., and Kosatky, T. (2010). Heat-related mortality: A review and exploration of heterogeneity. *Journal of Epidemiology and Community Health 64*(9), 753–760. https://doi.org/10.1136/jech.2009.087999

Harlan, S. L., Brazel, A. J., Prashad, L., Stefanov, W. L., and Larsen, L. (2006). Neighborhood microclimates and vulnerability to heat stress. *Social Science & Medicine 63*(11), 2847–2863.

Hester, E. T., and Bauman, K. S. (2013). Stream and retention pond thermal response to heated summer runoff from urban impervious surfaces. *Journal of the American Water Resources Association 49*(2), 328–342. https://doi.org/10.1111/jawr.12019

Hulme, M. (2014). Attributing weather extremes to 'climate change': A review. *Progress in Physical Geography 38*(4), 499–511. https://doi.org/10.1177/0309133314538644

Kotharkar, R., and Surawar, M. (2016). Land use, land cover, and population density impact on the formation of canopy urban heat islands through traverse survey in the Nagpur urban area, India. *Journal of Urban Planning and Development 142*(1), 04015003.

Liu, H., and Weng, Q. (2008). Seasonal variations in the relationship between landscape pattern and land surface temperature in Indianapolis, USA. *Environmental Monitoring and Assessment 144*(1–3), 199–219. https://doi.org/10.1007/s10661-007-9979-5

Liu, H., and Weng, Q. (2012). Enhancing temporal resolution of satellite imagery for public health studies: A case study of West Nile Virus outbreak in Los Angeles in 2007. *Remote Sensing of Environment 117*, 57–71. https://doi.org/10.1016/j.rse.2011.06.023

Liu, L., and Zhang, Y. (2011). Urban heat island analysis using the landsat TM data and ASTER Data: A case study in Hong Kong. *Remote Sensing 3*(7), 1535–1552. https://doi.org/10.3390/rs3071535

Lokhande, Y., Ghatak, S., Ghosh, D., and Pahwa, D. (2015). *MangingPeri-Urban Expansion: Knowlege Support to PEARL Programme under JNNURM.* Delhi, Beijing, New York: Nationa Institute of Urban Affairs (Cities Alliance: Cities without Slum).

Loughnan, M. E., Tapper, N. J., Lynch, K., McInnes, J., and Phan, T. (2012). *A Spatial Vulnerability Analysis of Urban Populations During Extreme Heat Events in Australian Capital Cities*. National Climate Change Adaptation Research Facility.

Mallick, J. (2014). Land characterization analysis of surface temperature of semi-arid mountainous City Abha, Saudi Arabia using remote sensing and GIS. *Journal of Geographic Information System 06*(06), 664–676. https://doi.org/10.4236/jgis.2014.66055

Meze-Hausken, E. (2008). On the (im-) possibilities of defining human climate thresholds. *Climatic Change 89*(3), 299–324.

Michelozzi, P., Accetta, G., De Sario, M., D'Ippoliti, D., Marino, C., Baccini, M., ... and Perucci, C. A. (2009). High temperature and hospitalizations for cardiovascular and respiratory causes in 12 European cities. *American Journal of Respiratory and Critical Care Medicine 179*(5), 383–389.

Mullins, J. T., and White, C. (2019). Temperature and mental health: Evidence from the spectrum of mental health outcomes. *Journal of Health Economics 68*, 102240.

Murage, P., Hajat, S., and Kovats, R. S. (2017). Effect of night-time temperatures on cause and age-specific mortality in London. *Environmental Epidemiology (Philadelphia, Pa.) 1*(2), e005.

Oke, T. R., and Maxwell, G. B. (1975). Urban heat island dynamics in Montreal and Vancouver. *Atmospheric Environment (1967) 9*(2), 191–200.

Purevdorj, T. S., Tateishi, R., Ishiyama, T., and Honda, Y. (1998). Relationships between percent vegetation cover and vegetation indices. *International Journal of Remote Sensing 19*(18), 3519–3535. https://doi.org/10.1080/014311698213795

Qin, Z., Karnieli, A., and Berliner, P. (2001). A mono-window algorithm for retrieving land surface temperature from Landsat TM data and its application to the Israel-Egypt border region. *International Journal of Remote Sensing 22*(18), 3719–3746. https://doi.org/10.1080/01431160010006971

Rifkin, D. I., Long, M. W., and Perry, M. J. (2018). Climate change and sleep: A systematic review of the literature and conceptual framework. *Sleep Medicine Reviews 42*, 3–9. https://doi.org/10.1016/j.smrv.2018.07.007

Rizwan, A. M., Dennis, L. Y., and Chunho, L. I. U. (2008). A review on the generation, determination and mitigation of Urban Heat Island. *Journal of Environmental Sciences 20*(1), 120–128.

Rosenfeld, A. H., Akbari, H., Romm, J. J., and Pomerantz, M. (1998). Cool communities: Strategies for heat island mitigation and smog reduction. *Energy and Buildings 28*(1), 51–62.

Rozenstein, O., Qin, Z., Derimian, Y., and Karnieli, A. (2014). Derivation of land surface temperature for landsat-8 TIRS using a split window algorithm. *Sensors (Switzerland) 14*(4), 5768–5780. https://doi.org/10.3390/s140405768

Shahmohamadi, P., Che-Ani, A. I., Etessam, I., Maulud, K. N. A., and Tawil, N. M. (2011). Healthy environment: The need to mitigate urban heat island effects on human health. *Procedia Engineering 20*, 61–70. https://doi.org/10.1016/j.proeng.2011.11.139

Sherwood, S. C., and Huber, M. (2010). An adaptability limit to climate change due to heat stress. *Proceedings of the National Academy of Sciences of the United States of America 107*(21), 9552–9555. https://doi.org/10.1073/pnas.0913352107

Shriram, M., (2017). *Urbanisation in the National Capital Region: Overcoming Challenges to Improve Live-ability*. India: Confederation of Indian Industry.

Singh, P., Kikon, N., and Verma, P. (2017). Impact of land use change and urbanization on urban heat island in Lucknow city, Central India. A remote sensing based estimate. *Sustainable Cities and Society 32*, 100–114. https://doi.org/10.1016/j.scs.2017.02.018

Sobrino, J. A., Jiménez-Muñoz, J. C., and Paolini, L. (2004). Land surface temperature retrieval from LANDSAT TM 5. *Remote Sensing of Environment 90*(4), 434–440. https://doi.org/10.1016/j.rse.2004.02.003

Song, Y., and Wu, C. (2018). Examining human heat stress with remote sensing technology. *GIScience& Remote Sensing 55*(1), 19–37.

Sung, T. I., Wu, P. C., Lung, S. C., Lin, C. Y., Chen, M. J., and Su, H. J. (2013). Relationship between heat index and mortality of 6 major cities in Taiwan. *Science of the Total Environment 442*, 275–281.

Taha, H. (1996). Modeling impacts of increased urban vegetation on ozone air quality in the South Coast Air Basin. *Atmospheric Environment 30*(20), 3423–3430.

Tan, J., Zheng, Y., Tang, X., Guo, C., Li, L., Song, G., Zhen, X., Yuan, D., Kalkstein, A. J., Li, F., Chen, H. (2010). The urban heat island and its impact on heat waves and human health in Shanghai. *International Journal of Biometeorology 54*(1), 75–84. https://doi.org/10.1007/s00484-009-0256-x

Thosar, S. S., Butler, M. P., and Shea, S. A. (2018). Role of the circadian system in cardiovascular disease. *The Journal of Clinical Investigation, 128*(6), 2157–2167.

Tomlinson, C. J., Chapman, L., Thornes, J. E., and Baker, C. J. (2011). Including the urban heat island in spatial heat health risk assessment strategies: A case study for Birmingham, UK. *International Journal of Health Geographics 10*(1), 1–14.

Tucker, C. J. (1979). Red and photographic infrared linear combinations for monitoring vegetation. *Remote sensing of Environment 8*(2), 127–150.

Voogt, J. A., and Oke, T. R. (2003). Thermal remote sensing of urban climates. *Remote Sensing of Environment, 86*(3), 370–384.

Wallace, R. F., Kriebel, D., Punnett, L., Wegman, D. H., Wenger, C. B., Gardner, J. W., and Gonzalez, R. R. (2005). The effects of continuous hot weather training on risk of exertional heat illness. *Medicine & Science in Sports & Exercise 37*(1), 84–90.

Wang, F., Qin, Z., Song, C., Tu, L., Karnieli, A., and Zhao, S. (2015). An improved mono-window algorithm for land surface temperature retrieval from landsat 8 thermal infrared sensor data. *Remote Sensing 7*(4), 4268–4289. https://doi.org/10.3390/rs70404268

Weng, Q., Lu, D., and Schubring, J. (2004). Estimation of land surface temperature–vegetation abundance relationship for urban heat island studies. *Remote Sensing of Environment 89*(4), 467–483.

Yang, L., Qian, F., Song, D. X., and Zheng, K. J. (2016). Research on urban heat-island effect. *Procedia Engineering 169*, 11–18.

Zeng, Q., Li, G., Cui, Y., Jiang, G., and Pan, X. (2016). Estimating temperature-mortality exposure-response relationships and optimum ambient temperature at the multi-city level of China. *International Journal of Environmental Research and Public Health 13*(3), 279.

Zha, Y., Gao, J., and Ni, S. (2003). Use of normalized difference built-up index in automatically mapping urban areas from TM imagery. *International Journal of Remote Sensing 24*(3), 583–594. https://doi.org/10.1080/01431160304987

Zhang, Y. D., Zhang, X. H., and Liu, S. R. (2011). Correlation analysis on normalized difference vegetation index (NDVI) of different vegetations and climatic factors in Southwest China. *Ying Yong Sheng Tai Xue Bao 22*(2), 323–330.

CHAPTER 11

A MULTIATTRIBUTE OPTIMIZATION MODELING OF SOCIOECONOMIC ISSUES OF SUSTAINABLE DEVELOPMENT GOALS IN INDIA: MISSION 2030

JAHANGIR CHAUHAN[1], AQSA ANJUM[1], U. M. MODIBBO[2], and IRFAN ALI[3]

[1]*Department of Commerce, Aligarh Muslim University, Aligarh, Uttar Pradesh, India*

[2]*Department of Operations Research, Modibbo Adama University Yola, Nigeria*

[3]*Department of Statistics & Operations Research, Aligarh Muslim University, Aligarh, Uttar Pradesh, India*

ABSTRACT

With massive exploitation of global reserves for natural resources and unhindered environmental degradation, there is an urgent need for sustainable developmental plans, to support the country's economic growth through selection and implementation of promising policies and viable programs for economic growth and development. For achieving sustainable development goals (SDGs), countries develop adequate strategies that simultaneously address efficiency in energy consumption, economic productivity, and reduction in greenhouse gas (GHG) emissions. There is a significant need to adopt the interdisciplinary approach to

Sustainable Development: A Geospatial and Statistical Perspective. Qazi Mazhar Ali, Rizwan Ahmad, Irfan Ali, & Prasenjit Chatterjee (Eds.)

DOI: 10.1201/9781779643247-11

address the economic and environmental problem through financial and computational modeling. Therefore, the purpose of the proposed study is to develop an efficient model and solution procedure to solve the problem under investigation with a sustainable approach. For this purpose, LINGO 17.0 was employed to enhance the relevance and reliability of the results. The proposed study, if implemented, will provide an insight into key decision-makers, economic, and financial analyst across different sectors of the economy, keeping in view the sustainable development goals mission 2030. A multiattribute optimization model for economic policies and SDGs has been employed in the Indian context, and the outcomes have discussed.

11.1 INTRODUCTION

Multiattribute Optimization Modeling is one of the operations research branches that deal with the modeling and the analysis of problematic real-world situation having several conflicting objectives which can be rightfully considered solution alternatives and trade-offs. The decision-maker chooses among various solutions in the feasible region, one that satisfies his criteria simultaneously without undermining both over achievements and under achievements of the targeted goals within the frame of interest. Over many years, different solution techniques have been developed and used to model and analyze multiattribute problems such as Multiattribute Utility Theory, Data Envelopment Analysis, Technique for Order of Preference by similarity to Ideal Solution, Analytical Hierarchy Process, Preference Ranking Organization Method for Enrichment of Evaluations, VIKOR, Elimination and Choice Expressing Reality, and Goal Programming (GP). The nature of the problem under study determines the type of method(s) or techniques to be used, which widely varies depending on the application domain. GP is a typical approach known to handle problems that have desired and targeted goals or aspirations. GP is flexible in terms of solving decision-making problem having competing objectives by way of minimizing deviations among the targeted and achievement values for potential trade-offs exploration. There exist variants of GP which include but not limited to lexicographic GP, weighted GP, polynomial GP, among others. Once a goal is deterministic and precisely defined, the standard GP model can be used to solve the problem. Generally, GP models are applied widely in the field of agriculture, commerce, economics, energy, environment, finance, health, manufacturing, resources planning, supply chain, sustainability to mention but a few. These are referred to in the studies conducted by Ragkos and Psychoudakis (2009), Prišenk et al. (2014), Aouni et al. (2014), Zopounidis et al. (2015), Zografidou et al. (2016), Nomani et al. (2017), Jones and Wall (2015), Jayaraman et al. (2017), Brandenburg (2015), Yu et al. (2015), Leung and Chan (2009), Ben Abdelaziz and Masmoudi (2012), Choudhary and Shankar (2014), and Gupta et al. (2018).

11.2 SUSTAINABLE DEVELOPMENT GOALS (SDGS) OF INDIA

Sustainability agenda of India requires sustaining economic growth, maximizing electricity consumptions, and minimizing GHG emissions with optimal human resources requirement in various sectors of the economy. This chapter addresses some valuable contributing factors to achieving the year 2030 agenda of SDGs of India. India is the second-largest most populous nation in the world after China with more than 1.3 billion citizens (17% of the population in the world), and it covers 6.4% of the world's landmark. More than 1.77 million citizens are homeless, which translate into 30% of its population living in poverty, and a large portion of its citizenries has no access to efficient electricity (Gupta et al., 2018).

According to BP (2018a) and World Bank (2017), the world's number three largest generator of electricity after China and USA is India with electricity per capita consumption of 1117 kWh that is less than one-third average consumption of the world and one-fourth of China. Unemployment is very high, about 8.3% of the population. The nominal GDP per capita is lower than many developing countries in the world, amounting to USD 1408 annually, with only 6.8% growth. Our natural environment needs to improve and preserve by every citizen in every nation of the world as a duty-bound. In India, it is a law according to Article 48A (Directictives Principles of State Policy) and Article 51A(g) (Fundamental Duty; MLJ, 2016a). Despite international commitments to reduce emissions to their barest minimum, GHG emissions in India accumulate and grow as a result of the climate change problem; it, therefore, pledges to minimize by 33–35% the intensity emissions of its GDP in the year 2030. India has counted among the committee of nations having one of the fastest-growing markets in the energy sector in the world, and projected by 2035 to become the second-largest energy demand contributor globally. The per capita energy consumption of India is 0.527 million tones (22.042 GJ), giving rise to 18% increase in global energy consumption. In terms of renewable energy, it has the fifth-largest world's wind power market, and determine to increase the solar power capacity to 100,000 MW by 2030 (Srikanth, 2018).

The country has a target of increasing the overall electricity generation capacity by 4.8% through a contribution of nuclear power within 25 years. Asian Development Bank report estimated the approximate cost of climate change adaptation for India in the energy sector as USD 7.7 billion (at 2014–2015 prices) by the year 2030 while the economic damage and losses resulting from climate change are estimated to be 1.8% of the annual GDP by 2050 (Gupta et al., 2018). Modern energy, reliable and affordable services, are of paramount importance to SDGs achievement for India. Goal 7 of the SDG is central to the majority of challenges of India, which must be overcome to guarantee future sustainability. In this regard, this chapter seeks to examine India's action plans for achieving SDG 7 as well as the Nationally Determined Contributions by 2030. We consider the following socioeconomic indicators and their attributes in our model for optimizing the mission 2030 sustainable development of India (see Table 11.1).

TABLE 11.1 Identified Goals for the Year 2030

Economic Indicator	GDP (in Billion INR)	ECC (in GWh)	GHG (in Million Tones)	Number of Employment ('000)
Value	98,576.73	890,055	1904.73	462,410
Goal by year 2030	367,935.59	2,684,750.195	5700	742,032
Annual growth rate	8.58%	6.71%	4.88%	3.0%

Source: UNFCCC, 2015.

According to CEA (2018a), out of 1309 TWh total electricity generated by utilities, gas, coal, nuclear, hydro, and renewable energy sources such as solar, wind, and bio-energy contributed 4%, 75%, 3%, 10%, and 8%, respectively, in the 2017–2018 financial year. Additionally, captive power plants also generated 19.7 TWh during that year (MOSPI, 2018).

11.3 SDGS MODEL FORMULATION

In a real-life situation, most optimization problems have enormous conflicting attributes, imprecise goals, estimates, and criteria. Thus, GP approach offers analytical means of modeling and solving such issues. We develop a GP model that simultaneously optimize rules for mission 2030 of India relating to electricity consumption, economic growth, GHG emissions, and workforce allocation (see Table 11.1). We validated the model with data from nine sectors of Indian economy (see Table 11.2).

TABLE 11.2 Sectorwise Economic Contribution.

S/N	Economic Sector	GDP (Billion INR)[a]	ECC (GWh)[b]	GHG (Million Tonnes)[b]	Number of Employment ('000)[c]
1	Agriculture, forestry, and fishing	15,968.77	144,189	334.41	23,430
2	Mining and quarrying	2826.05	25,544.6	165.31	18,770
3	Manufacturing	17,702.32	159,854	117.32	110,670
4	Electricity, gas, water supply, and other utility services	2322.69	21,005.3	719.30	13,770
5	Constructions	7926.63	71,560.4	129.92	14,580
6	Trade, repair, hotels, and restaurants	12,110.97	109,388	100.87	12,210
7	Transport, storage, communication, and services related to broadcasting	6541.33	59,099.7	142.04	39,290
8	Financial, real estate, and professional services	20,557.06	185,576	57.73	50,900
9	Community, social, and personal services	12,620.91	113,838	137.84	178,710

[a]Planning Commission, Government of India.

[b]INDC Report submitted to UNFCCC.

[c]Ministry of Labour & Employment, Director General of Employment and Training, India.

11.3.1 SDGS MODEL

$$F_1(X) = \sum_{i=1}^{9} \left[\frac{\text{GDP}_j}{E_i} \right] x_j \tag{11.1}$$

$$F_2(X) = \sum_{i=1}^{9} \left[\frac{(\text{ECC})_j}{E_i} \right] x_j \tag{11.2}$$

$$F_3(X) = \sum_{i=1}^{9} \left[\frac{(\text{GHG})_j}{E_i} \right] x_j \tag{11.3}$$

$$F_2(X) = \sum_{i=1}^{9} x_j \tag{11.4}$$

Subject to the set of constraints:

$$\sum_{i=1}^{9} \left[\frac{\text{GDP}_j}{E_i} \right] x_j \geq (\text{GDP})_g \tag{11.5}$$

$$\sum_{i=1}^{9} \left[\frac{(\text{ECC})_j}{E_i} \right] x_j \geq (\text{ECC})_g \tag{11.6}$$

$$\sum_{i=1}^{9} \left[\frac{(\text{GHG})_j}{E_i} \right] x_j \leq (\text{GHG})_g \tag{11.7}$$

$$\sum_{i=1}^{9} x_j \leq E_g \tag{11.8}$$

$$E_j \leq x_j \leq E_g \; j = 1,2,\ldots,9 \tag{11.9}$$

where

- F_1 is the goal function that seeks to optimize the GDP per capita across the *j*th economic sector.
- F_2 is the goal function that seeks to optimize the per capita electricity consumption (ECC) across the *j*th economic sector.
- F_3 is the goal function that seeks to minimize the per capita GHG emission across all the *j*th economic sector.
- F_4 is the goal function that seeks to optimize the number of employees across the *j*th economic sector.
- Constraint (11.5) places restrictions on the GDP goal.
- Constraint (11.6) places limitations on the ECCgoal.
- Constraint (11.7) imposes restrictions on the goal of GHG emission.
- Constraint (11.8) imposes restrictions on the goal of employment.
- x_j is the employees in the *j*th contributing sector of the economy.
- E_j is the employment in the *j*th economic sector.
- $(\text{ECC})_j$ is the electricity consumption in the *j*th economic industry.
- $(\text{GHG})_j$ is the green house gas emission in the *j*th contributing sector.

- $(GDP)_j$ is the gross domestic product in the *j*th sector.
- $(GHG)_g$ is the green house gas emission goal for SDG.
- $(GDP)_g$ is the gross domestic product goal for SDG.
- $(ECC)_g$ is the electricity consumption goal for SDG.
- E_g is the employment goal at the annual growth rate.

11.4 GOAL PROGRAMMING PROCEDURE FOR SOLVING SDGS MODEL

Charnes and Cooper (1961, 1977) introduced the concept of GP and suggested a procedure to solve linear programming problems having goals that are conflicting in nature. Ijiri (1965) coined out the preemptive priority concept where incommensurable goals are assigned priority with different weights at the same priority level. Decision-making always involves competing for interest, among which priority is the sort to be ascribed to the goals. The decision-maker seeks to balance among the targets of interest, as such GP give a compromise solution to all the objectives rather than the single optimum solution. The overall idea of GP is to obtain a solution that will satisfy all the limitations and minimize their possible deviations at the same time. These deviations can be that of overachievement or underachievement of the goals; in some situations, both deviations from the targets are acceptable to the decision-maker at a certain aspiration level. While in other cases, the variations weighted to relatively reflect the importance of the goals.

The procedures are as follows:

$$\text{Optimize } f_i(X) = \sum_{j=1}^{n} C_{ij} x_j \ , i = 1,2,\ldots,m \tag{11.10}$$

where $X = (x_1, x_2, x_3 \ldots, x_n)$

subject to :

$$\sum_{j=1}^{n} C_{ij} x_j \geq L_i \tag{11.11}$$

$$\sum_{j=1}^{n} C_{ij} x_j \leq U_i \tag{11.12}$$

$$\sum_{j=1}^{n} C_{ij} x_j = G_i \tag{11.13}$$

Equation 11.10 is the objective function that seeks to evaluate specific effectiveness measures. Equation 11.11, is the *i*th objective function, a lower-valued goal, which the decision-maker does not want to fall below. In order words, the DM desires to achieve a minimum value of L_i meaning that exceeding L_i value is allowed. Equation 11.12, is the *i*th objective function, an upper-valued goal, which the decision-maker does not want to exceed above. In order words, the DM desires to achieve a maximum possible value of U_i meaning that exceeding U_i value is allowed to a certain level of tolerance. Equation 11.13, is the *i*th objective function, a specific goal, which the decision-maker targeted to achieve. In order words, the DM specified the desired goal, which can deviate from either direction (under- and overachievements). Equations 11.11 and 11.12 are a soft and flexible constraint and, hence, written as

$$L_i \leq f_i(X) \leq U_i \tag{11.14}$$

Let us define the deviations that are central to the construct of GP as follows:

- $\boldsymbol{D}_{ij}^{+}$ denotes the overachievement value of *i*th objective or the amount by which the *i*th objective exceeded the targeted goal, that is, a positive deviation.
- $\boldsymbol{D}_{ij}^{-}$ denotes the underachievement value of *i*th objective or the amount by which the *i*th objective fall below the targeted goal, that is, a negative deviation.

Equations 11.11–11.13 can be rewritten in the standard form to measure the deviations as follows:

$$\sum_{j=1}^{n} C_{ij}x_j - \boldsymbol{D}_{ij}^{-} + \boldsymbol{D}_{ij}^{+} = L_i \tag{11.15}$$

$$\sum_{j=1}^{n} C_{ij}x_j - \boldsymbol{D}_{ij}^{-} + \boldsymbol{D}_{ij}^{+} = U_i \tag{11.16}$$

$$\sum_{j=1}^{n} C_{ij}x_j - \boldsymbol{D}_{ij}^{-} + \boldsymbol{D}_{ij}^{+} = G_i \tag{11.17}$$

These deviations as appeared in the constraints 11.15–11.17 are then penalized in the objective functions by the penalty weights $\boldsymbol{P}_i^{+}$ and $\boldsymbol{P}_i^{-}$, where $\boldsymbol{P}_i^{-}$ represent the underachievement of goal penalty and $\boldsymbol{P}_i^{+}$ represents the overachievement of goal penalty. Thus, we have the following penalty association of goals:

For L_i goal-type: $\boldsymbol{P}_i^- > 0, \boldsymbol{P}_i^+ = 0;$ (11.18)

for U_i goal-type: $\boldsymbol{P}_i^- = 0, \boldsymbol{P}_i^+ > 0;$ (11.19)

for G_i goal-type: $\boldsymbol{P}_i^- > 0, \boldsymbol{P}_i^+ = 0;$ (11.20)

Here, for instance, we introduce a hard constraint $\boldsymbol{P}_i^- > 0$ if we want to penalize the underachievement of say L_i also, maintain its achievement by setting $\boldsymbol{P}_i^+ = 0$. The same analogy goes to the rest of the goals. The compromised objective function of the GP called for minimizing the total deviations from their respective goals together with the associated weights. Thus, the complete formulation outlined below:

$$\text{Minimize } Z_i = \sum_{j=1}^{n} (\boldsymbol{P}_i^+ \boldsymbol{D}_{ij}^+ + \boldsymbol{P}_i^- \boldsymbol{D}_{ij}^-),\ \boldsymbol{i} = 1, 2, \ldots, \boldsymbol{m} \quad (11.21)$$

Subject to the constraints 11.15–11.20.

11.5 MATHEMATICAL MODELING OF INDIA'S SDGS

Data in Tables 11.1 and 11.2 in Section 11.2 are used to construct the following mathematical expressions using the SDGs model in Section 11.3.1. Thus, we have the following equations:

$$F_1(X) = 0.681523x_1 + 0.1505621x_2 + 0.1599559x_3 + 0.1686776x_4 + 0.5436646x_5 + 0.9918894x_6 + 0.1664884x_7 + 0.4038715x_8 + 0.0706223x_9$$

$$F_2(X) = 6.15403x_1 + 1.36093x_2 + 1.44442x_3 + 1.52544x_4 + 4.90812x_5 + 8.95887x_6 + 1.50419x_7 + 3.6459x_8 + 0.63699x_9$$

$$F_3(X) = 0.0142727x_1 + 0.0088071x_2 + 0.0010601x_3 + 0.0522367x_4 + 0.0089108x_5 + 0.0082613x_6 + 0.0036152x_7 + 0.0011342x_8 + 0.0007713x_9$$

$$F_4(X) = x_1 + x_2 + x_3 + x_4 + x_5 + x_6 + x_7 + x_8 + x_9$$

Subject to:

$$0.681523x_1 + 0.1505621x_2 + 0.1599559x_3 + 0.1686776x_4 + 0.5436646x_5 + 0.9918894x_6 + 0.1664884x_7 + 0.4038715x_8 + 0.0706223x_9 \geq 367{,}935.59;$$

$$6.15403x_1 + 1.36093x_2 + 1.44442x_3 + 1.52544x_4 + 4.90812x_5 + 8.95887x_6 + 1.50419x_7 + 3.6459x_8 + 0.63699x_9 \geq 2{,}684{,}750.195;$$

$$0.0142727x_1 + 0.0088071x_2 + 0.0010601x_3 + 0.0522367x_4 + 0.0089108x_5 + 0.0082613x_6 + 0.0036152x_7 + 0.0011342x_8 + 0.0007713x_9 \leq 5{,}700{,}000;$$

$$x_1 + x_2 + x_3 + x_4 + x_5 + x_6 + x_7 + x_8 + x_9 \geq 742{,}032;$$

$$x_1 \geq 23{,}430;\ x_2 \geq 18{,}770;\ x_3 \geq 110{,}670;\ x_4 \geq 13{,}770;\ x_5 \geq 14{,}580;$$
$$x_6 \geq 12{,}210;\ x_7 \geq 39{,}290;\ x_8 \geq 50{,}900;\ x_9 \geq 178{,}710;$$
$$D_{ij} \geq 0 \text{ and integer, } i = j = 1,2, \ldots, 9;$$

The mathematical model in Section 11.5 is converted using GP procedures in Section 11.4, and LINGO Optimization package 17.0 was employed to obtain the result, as shown in Table 11.3 in Section 11.6, respectively.

11.5.1 SDGS MODEL IN GP STANDARD FORM

$$\text{Min } Z = 0.25*(D_{11} + D_{12}) + 0.25*(D_{21} + D_{22}) + 0.25*(D_{31} + D_{32}) + 0.25*(D_{41} + D_{42})$$

$$0.681523*x_1 + 0.1505621*x_2 + 0.1599559*x_3 + 0.1686776*x_4 + 0.5436646*x_5 + 0.9918894*x_6 + 0.1664884*x_7 + 0.4038715*x_8 + 0.0706223*x_9 + D_{11} - D_{12} = 367935.59;$$

$$6.15403*x_1 + 1.36093*x_2 + 1.44442*x_3 + 1.52544*x_4 + 4.90812*x_5 + 8.95887*x_6 + 1.50419*x_7 + 3.6459*x_8 + 0.63699*x_9 + D_{21} - D_{22} = 2684750.195;$$

$$0.0142727*x_1 + 0.0088071*x_2 + 0.0010601*x_3 + 0.0522367*x_4 + 0.0089108*x_5 + 0.0082613*x_6 + 0.0036152*x_7 + 0.0011342*x_8 + 0.0007713*x_9 + D_{31} - D_{32} = 5{,}700{,}000;$$

$$x_1 + x_2 + x_3 + x_4 + x_5 + x_6 + x_7 + x_8 + x_9 + D_{41} - D_{42} = 742{,}032;$$

Constraints:

$x_1 \geq 23{,}430;$

$x_2 \geq 18{,}770;$

$x_3 110{,}670;$

$x_4 \geq 13{,}770;$

$x_5 14{,}580;$

$x_6 \geq 12{,}210;$

$x_7 \geq 39{,}290;$

$x_8 \geq 50{,}900;$

$x_9 \geq 178{,}710;$

$D_{ij} \geq 0$ and integer, $i = j = 1, 2, \ldots, 9.$

11.6 SOLUTION

In this section, we employed the GP technique for minimizing the under-achievement and overachievement of the goals in model 2 above. We then solved the model by LINGO optimization software version 17.0 and obtained the solution as in Table 11.3. The result indicated a significant positive deviation in D_{31} and D_{11}. We can interpret these values like that; a large portion of energy and GDP are required to satisfy all the four attributes simultaneously and to be able to reduce the GHG emission in India by 2030. It means that the sustainable goals mission 2030 of India can only be achievable if other sources of renewable energy such as wind, nuclear, and solar incorporated as an alternative to the growing demands of energy in India.

TABLE 11.3 LINGO Solver Result Version 17.0

Deviations	Value	Reduced Cost	Variable	Value	Reduced Cost
D_{11}	**70678.44**	0.000000	x_1	23430.00	0.2500000
D_{12}	0.000000	0.5000000	x_2	18770.00	0.2620716
D_{21}	0.000000	0.2762072	x_3	110670.0	0.2581855
D_{22}	0.000000	0.2237928	x_4	109434.7	0.1107493
D_{31}	**5691578**	0.000000	x_5	14580.00	0.1285127
D_{32}	0.000000	0.5000000	x_6	196247.3	0.5735166
D_{41}	0.000000	0.2652511	x_7	39290.00	0.1214577
D_{42}	0.000000	0.2347489	x_8	50900.00	0.9548388
			x_9	178710.0	0.1409642

11.7 CONCLUSION

Multiattributes optimization modeling of sustainability goals have a vital role in formulating policies and developmental issues. In this chapter, a multiattribute optimization model for economic policies and SDGs has been employed in the Indian context, integrating economic growth, ECC, GHG emissions, and employment by the year 2030. The outcomes have been discussed; the model provided mathematical justification for changes and adjustment in the energy sector. India will appreciate the introduction and incorporation of renewable energy in the next 10 years to come.

KEYWORDS

- **socioeconomic problems**
- **sustainable development goals**
- **optimization modeling**
- **sustainable and economic development**
- **socioeconomic policies**

REFERENCES

Aouni, B., Colapinto, C., and La Torre, D. (2014). Financial portfolio management through the goalprogramming model: Current state-of-the-art. *European Journal of Operational Research 234*(2), 536–545.

Ben Abdelaziz, F., and Masmoudi, M. (2012). A multiobjective stochastic program for hospital bed planning. *The Journal of the Operational Research Society 63*(4), 530–538.

BP. (2018a). BP statistical review of world energy—all data. Retrieved from https://www.bp.com/en/global/corporate/energy-economics/statistical-review-of-world energy/downloads. html (Accessed July 18, 2019).

Brandenburg, M. (2015). Low carbon supply chain configuration for a new product—a goal programming approach. *International Journal of Production Research 53*(21), 6588–6610.

CEA. (2018a). Executive summary for march. Retrieved from ⟨http://www.cea.nic.in/reports/monthly/executivesummary/2018/exe_summary-03.pdf⟩ (Last Accessed August 9, 2019).

Charnes, A., and Cooper, W. W. (1961). *Management Model and Industrial Application Oflinear Programming* (Vol. 1). New York: Wiley.

Charnes, A., and Cooper. W. W. (1977). Goal programmingand multiple objectives optimisations. *European Journal of Operational Research 1*, 39–54. doi:10.1016/S0377-2217(77)81007-2.

Choudhary, D., and Shankar, R. (2014). A goal programming model for joint decision making of inventory lotsize, supplier selection and carrier selection. *Computers & Industrial Engineering 71*, 1–9.

Gupta, S., Fügenschuh, A., and Ali, I. (2018). A multi-criteria goal programming model to analyze the sustainable goals of India. *Sustainability 10*(3), 778.

Ijiri, Y. (1965). Axioms and structures of conventional accounting measurement. *The Accounting Review 40*(1), 36.

Jayaraman, R., Colapinto, C., Liuzzi, D., and La Torre, D. (2017). Planning sustainable development through a scenario-based stochastic goal programming model. *Operational Research 17*(3), 789–805.

Leung, S. C., and Chan, S. S. (2009). A goal programming model for aggregate production planning with resource utilization constraint. *Computers & Industrial Engineering 56*(3), 1053–1064.

Ministry of Labour & Employment, Director General ofEmployment and Training https://data.gov.in/catalog/employment-organised-sectors-public-and-private.

MLJ. (2016a). (Ministry of Law and Justice). The Constitution of India. Retrieved from ⟨http://www.legislative.gov.in/sites/default/files/coi4March 2016.pdf⟩ (Last Accessed May 10, 2018).

MOSPI. (2018). (Ministry of Statistics & Program Implementation). Energy Statistics. retrieved From http://mospi.nic.in/sites/default/files/publication_reports/Energy_Statistics_2018.pd (Last Accessed 9, August 2019).

Nomani, M. A., Ali, I., Fügenschuh, A., and Ahmed, A. (2017). A fuzzy goal programming approach to analyse sustainable development goals of India. *Applied Economics Letters 24*(7), 443–447.

Prišenk, J., Turk, J., Rozman, Č., Borec, A., Zrakić, M., and Pažek, K. (2014). Advantages of combining linear programming and weighted goal programming for agriculture application. *Operational Research 14*(2), 253–260.

Ragkos, A., and Psychoudakis, A. (2009). Minimizing adverse environmental effects of agriculture: A multi-objective programming approach. *Operational Research 9*(3), 267–280.

Srikanth, R. (2018). India's sustainable development goals–Glide path for India's power sector. *Energy Policy 123*, 325–336.

United Nations Framework Convention on Climate Change. (2015). http://www4.unfccc.int/submissions/INDC/Published%20Documents/India/1/INDIA%20INDC%20TO%20UNFCCC.pdf

World Bank. (2017). State of electricity access report. Retrieved from http://documents. World bank.org/curated/en/364571494517675149/pdf/114841-REVISED-JUNE12-FINAL-SEAR-web-REV-optimized.pdf (Accessed August 6, 2019).

Yu, S., Zhang, S., Agbemabiese, L., and Zhang, F. (2015). Multi-stage goal programming models for production optimization in the middle and later periods of oilfield development. *Annals of Operations Research* https://doi.org/10.1007/s10479-015-1866-9

Zografidou, E., Petridis, K., Arabatzis, G., and Dey, P. K. (2016). Optimal design of the renewable energy map of greece using weighted goal-programming and data Envelopment analysis. *Computers & Operations Research 66*, 313–326. http://dx.doi.org/10.1016/j.cor.2015.03.012

CHAPTER 12

ENERGY SUPPLY COST SUSTAINABILITY: A CASE STUDY OF EDUCATIONAL INSTITUTION

AQSA ANJUM and JAHANGIR CHAUHAN

Department of Commerce, Aligarh Muslim University, Aligarh, Uttar Pradesh, India

ABSTRACT

With a population of 1.2 billion and a GDP of 2622.98 billion dollars in 2019, India is one of the biggest and the fastest-growing economies in the world. India's energy demand has risen by more than 60% since 2000, yet significant disparities exist across different regions. Coal, oil, and solid biomass account for 80% of energy requirements. India ranks third in terms of carbon dioxide emission. Its electricity industry, particularly, has a carbon intensity that is significantly above the global average. The government of India has boosted the target of 370 to 450 GW by 2030. India's renewable capacity installation has reached 101.53 GW, in which installed solar power capacity is 47.7 GW. Educational institutions play a vital role in creating human capital. Still, the benefits of spending on these social sectors take time to reap.

This study examines whether shifting to solar power generation can reduce the overall cost in general and the electricity cost for educational institutions. For this purpose, a case study on Aligarh Muslim University (AMU) using solar power generators has been chosen. Among all Indian universities, AMU holds the biggest solar power plant, which was established

Sustainable Development: A Geospatial and Statistical Perspective. Qazi Mazhar Ali, Rizwan Ahmad, Irfan Ali, & Prasenjit Chatterjee (Eds.)

DOI: 10.1201/9781779643247-12

under the Solar Energy Corporation of India (SECI) initiative. The total solar capacity installed on the campus is 6.5 MW. It is also outfitted with a 3.3-MW ground-mounted solar plant. The study would be beneficial for educational institutions that are dependent on energy and to fulfill Goal 7 of Sustainable Development Goals (SDGs) by 2030.

12.1 INTRODUCTION

Two broad power sources are conventional and nonconventional. Traditional power sources, natural gas, coal, oil, and nuclear energy, result in emissions of CO_2 that negatively affect the atmosphere. Therefore, emphasis is put on renewable sources to foster economic development, tackle environmental destruction, and save the loss of natural capital. India has explored only a fraction of renewable energy's tremendous potential; thus, the target has been boosted from 175 to 450 GW of added capacity by 2030. India's renewable capacity installation reached 101.5 GW in September 2021. Solar is the most significant contributor, accounting for 46% of the overall renewable energy mix, followed by wind at 39%. Around 66 GW of solar, wind, and hybrid projects are currently in the queue, with most expected to be completed in the next 4–5 years. Solar energy is harnessed using various technologies, including radiant heat, photovoltaic (PV), solar thermal, and a few more. Solar photovoltaic (SPV) has more scope than other renewables due to India's availability of raw materials and electricity (sun) sources. Its fortunate position near the equator has enormous potential. Per year, it gets 3000 h of sunshine. The average solar radiation intensity is 200 MW per square km, as shown in Figure 12.1. India's geographical area is 3287 million square kilometers. However, only 413 million square kilometer can be utilized for solar energy installation. Though it was possible to use 10% of the area, the solar energy available will be 8 million MW. Solar energy collected per 1 m^2 of solar collector is equivalent to the energy released by burning 1 kg of coal. The economy is being pushed to accept renewable energy by several causes; increasing energy needs, lowering costs, a persistent energy shortfall, hefty state subsidies, and a desire for ecologically benign technologies are demand-side causes for solar expansion. Supply-side factors include overreliance on coal, oil importers, heavy subsidies for companies, sizeable rural demand, and abundant sun rays.

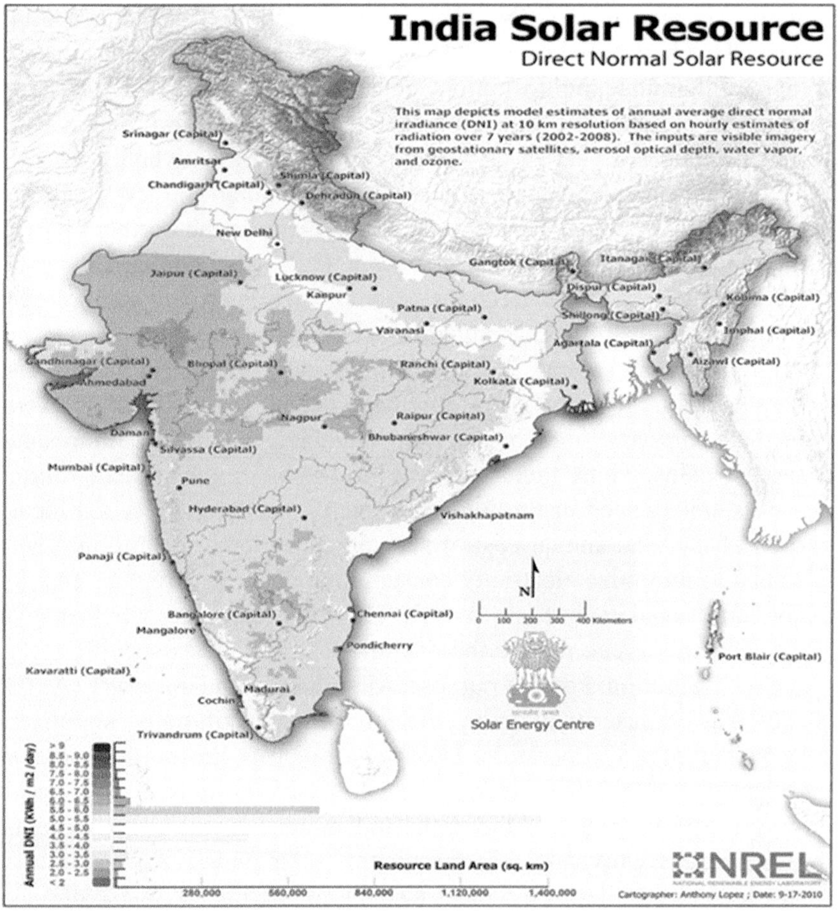

FIGURE 12.1 India Solar Resource data.

Source: NREL

There are around 1043 universities and about 42,000 colleges in India as per a survey of higher education in 2020. IIT Kharagpur requested a proposal for 5.5 MW of solar projects in May 2018. During the same time period, the Govind Ballabh Pant University in Uttarakhand bid on a 5-MW grid-connected rooftop SPV project. Amity University installed 1.8 MW of rooftop SPV systems throughout the campuses in July 2018. Jamia Milia Islamia University in Delhi established a 2.25-MW rooftop solar plant in November 2018. The IIT Indore campus joined the list of sustainable campuses in December 2018 when it installed a 422-kW solar rooftop system. Jawaharlal Nehru University erected a 500-kW rooftop solar system to illuminate the institution about the same period. In addition, St. Xavier's High School in Mumbai installed a 35-kW rooftop solar system, saving $70,000 per year in electricity costs. Tripura University recently installed rooftop solar panels on its campus, totaling 500 kW, scattered over multiple buildings. This is the first university in the northern part of India to build a large-scale PV power plant. O.P. Jindal Global University has become one of the universities for implementing green technologies on its campus in 2021. Educational institutions play a vital role in the creation of human capital. Still, the benefits of spending on these social sectors take time to reap. Therefore, the expenditure on education exerts immediate pressure on the government budget without immediate economic gains. Thus, there is a need to minimize the cost of routine expenses among such institutions. Electricity bills comprise a significant portion of the total annual cost of such an educational institution. There is a need to minimize electricity consumption charges, a regular cost for major educational institutions in India. This study examines whether solar power generation can reduce the overall electricity costs specific to the academic institution. For this purpose, Aligarh Muslim University (AMU) is chosen as the case study among the educational institutions" significant solar power energy generators. The Aligarh district lies on the banks of the Ganges and Yamuna rivers. The University campus is over 467.6 ha in the city of Aligarh. Among the Indian educational establishments, this prestigious University has the biggest solar power plant. The total solar capacity installed is 6.5 MW; also outfitted with a 3.3-MW ground-mounted solar plant.

12.2 THEORETICAL BACKGROUND

Solar energy offers more advantages than other renewable sources. Previously, solar energy was connected with household and commercial reasons, with minimal focus placed on the education sector in greening the globe. The Ministry of New and Renewable Energy launched several projects to construct green campuses to utilize institutes' backgrounds/space for renewable energy generation. Various studies have been done by researchers on the energy supply of educational institutions. Mendu et al. (2020) studied the integrated energy supply for educational institutions to check the feasibility of green energy deployment. The study concluded that a grid-connected solar system is the best option to meet the power needs. Baitule and Sudhakar (2017) assessed the feasibility of installing SPV in the academic campus at Maulana Azad National Institute of Technology in India. A SPV-based captive plant with a capacity of 5 MW was proposed. The performance of the proposed system was explored using PVsyst and System Advisor Model software. Ramana et al. (2015) studied the solar plant power output, cost production, and associated difficulties of employing a solar system in educational institutes. According to the results, if the cost of production of solar electricity is equivalent to the conventional power plant, solar power will become the primary energy source. Bhuyan et al. (2017) examined the performance of a solar power plant in on-grid and off-grid mode for meeting load demand and the economic benefits of the Micro-Electro-Mechanical Systems institute. The study explained the economic benefit of a solar plant in reducing the monthly grid unit use. In the case of an off-grid solar plant, reducing diesel consumption would provide economic benefits.

Furthermore, Bano and Rao (2016) analyzed the effect of lower module prices on the Levelized Cost of Electricity of a 1-MW SPV generating station. Nagar and Gidwani (2018) assessed the solar power plant's performance ratio and cumulative utilization factor to analyze solar power generation feasibility. The monthly performance ratio was found to be differed by 82–89% from the value assumed after calculation. Khan et al. (2021) study focused on effective energy management and estimating solar energy that can be generated based on an audit completed on residences of an Indian university. The study estimated that 6.3 crore can be saved each year if the recommended adjustments are adopted throughout all residence halls. Tariq et al. (2021) investigated that dust deposition on PV module surfaces leads to decreased module efficiency, caused by a reduction in PV module output power and solar irradiation. Rupal et al. (2016) presented a case study of a northern Indian academic institution for significant reductions in energy consumption. The study described an analysis that resulted in a 35% reduction in energy usage and a 177-ton reduction in carbon footprint in an Indian institute. The study confirmed that renewable energy sources might cover 25% of the energy needs of institute buildings and hostels after retrofitting procedures and installing a solar rooftop power plant. Ahmad and Alam (2017b) explored the energy utilization trends for analyzing, designing, and implementing an appropriate microgrid deployment strategy in an Indian Institute. The results showed that the total system cost under grid plus SPV mode is 494.92 million, saving approximately INR 864.2 million. The suggested topology will mitigate a significant amount of greenhouse gas emissions. Ansari et al. (2021) estimated the net present cost, cost of electricity, and pollution control using renewable energy technologies. A comparison between diesel-only systems and hybrid systems was made in the study. The study found the best alternatives for combining SPV with diesel generators having 22,368,262 NPV and 0.262 $/kWh levelized electricity cost. The study of Ahmad and Iqbal (2012) focused on energy saving to mitigate the electricity crisis of an Indian academic institute. The study estimated that using renewable sources of electricity instead of Dakshinanchal Vidyut Vitran Nigam Ltd.-Uttar Pradesh Power Corporation Ltd. (DVVNL-UPPCL) saves INR 3456.25 per hour. Solar energy is not limited to a particular sector. For its expansion, all sectors need to come together. Werulkar and Kulkarni (2015) examined the efficiency of a household SPV system. The study findings stated that the existing PV system with utility backup could save energy.

12.3 SOLAR PLANT OF AMU

The university is the leading academic institution in implementing green and renewable energy sources in more than 35 places. The campus has an installed capacity of 13.63 MVA, in which the rooftop installed capacity is 3.124 MWp. The rooftop project has been established under the SECI program. Twenty-two percent of the consumption is by residential halls, staff quarters, and communication towers, including line losses. Offices, departments, and street lighting consume 66% of the total electricity consumption. Before solar, AMU purchased electricity from DVVNL-UPPCL, Uttar Pradesh (MERCOM India, 2019). UPPCL had given the AMU campus a load of 6.1 MVA.

However, in 2013, university usage exceeded the limit of 7.952 MVA, and in 2015, it exceeded the limit of 8.176 MVA (Ahmad and Alam, 2017). This excess demand was owing to an increase in load demand. Load shedding is not an issue on campus, as a single-point supply serves the campus at 33 kV. The university's yearly electricity consumption is estimated to be over 2.6 million units, requiring the university's budget. As a result, the university administration decided to mount a grid-connected solar plant on campus. Simulation tests were conducted to determine the plant's optimal capacity before starting the installation. According to simulation studies, the facility generates roughly 45 lakh power units annually, with a payback period of about six years. The electricity generated by the project is fed into the university network via a transmission line that runs from the solar farm to the university polytechnic's 33-kV specialized substation, roughly 2.5 km away (Ahmad and Alam, 2017). It comprises 20 crore projects for a 3.3-MWp solar farm and is the first of its kind in any academic institute. The site has an extensive electricity network with a capacity of 8.5 MW and an annual expenditure of over INR 30 crores. A 4.8-MW capacity project has been functioning on campus for the past two years. A 3.3-MW ground-mounted solar plant is also available at the campus (MERCOM India, 2019).

12.4 METHODOLOGY

Quarterly electricity bills were used to acquire the energy usage data, collected from AMU's electricity department. The data of each building block's capacity was gathered. Following that, energy consumption data from UPPCL DVNL is used. Besides that, energy generation, energy consumption, and installed capacity of the solar plant were taken. Then, the bills related to the amount paid to CleanMax for solar and the amount paid to UPPCL were taken. Finally, the total cost paid per unit, savings due to solar power generation, and the growth potential of solar were estimated.

12.5 RESULTS AND DISCUSSION

Solar energy is being harnessed based on a complete audit of a hall of residences, offices, and departments to meet the demands stably of the institution. Year-wise rooftop installed capacity is shown in Figure 12.2. Rooftop capacity has increased from 1455.8 to 3474.25 kWh. Solar power generation from rooftop plants and ground-mounted plants has been shown in Figure 12.3. Electricity generation from ground-mounted plants has increased from 3,134,652 to 4,151,330 kWh in 2017 and electricity generation from rooftop has been found fluctuating since consumption got stagnant since hostels were closed during a pandemic. Energy consumption is provided in Figure 12.4, indicating increased annual energy consumption; then, a fall in energy consumption can be seen due to COVID-19. The total capacity of the buildings of the institute is 3124 KWp. The amount paid to CleanMax for solar and UPPCL has been shown in Figure 12.5, indicating the reduction in cost paid due to solar contribution. The study estimated that the per unit cost of solar power generation is reducing, leading to a reduction in expenses of the institution. The growth potential of solar shows the prospects of solar generation in upcoming years, as shown in Table 12.1. One of the benefits of solar energy for educational institutions is that most work is done during the day, especially in departments and offices. Thus, it is confirmed that installing a solar power plant would generate significant savings, which can be invested in other areas of development, as illustrated in Table 12.1. The solar plant mitigates 2069 tons of carbon emissions. Besides monetary savings, it also contributes to mitigating climate change and reducing the burden on UPPCL.

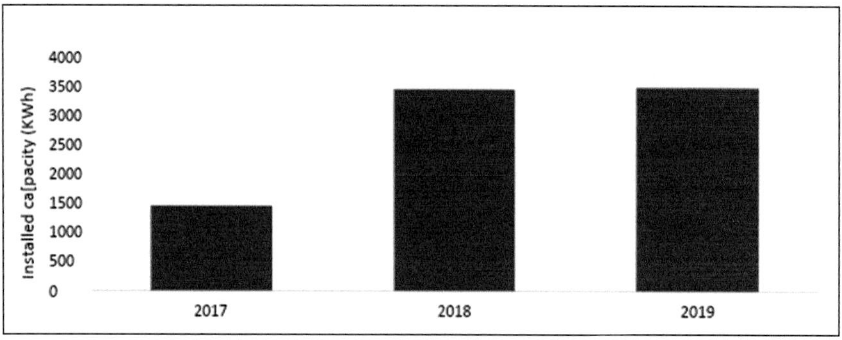

FIGURE 12.2 Annual rooftop installed capacity.

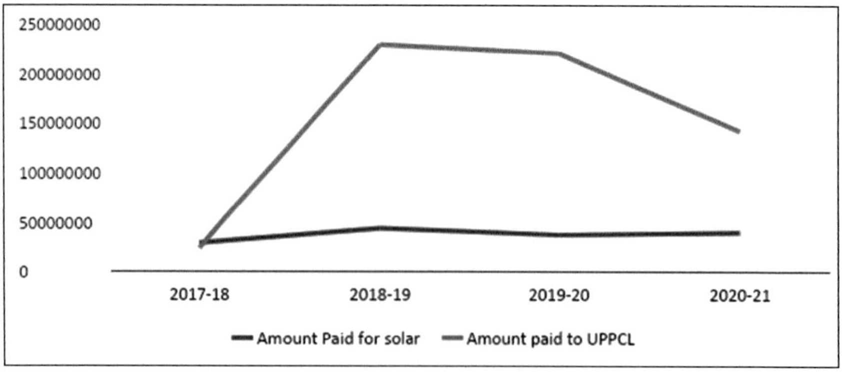

FIGURE 12.3 Cost paid by the institution.

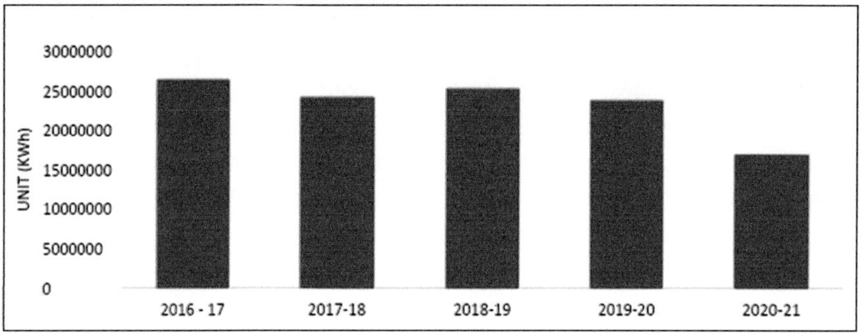

FIGURE 12.4 Energy consumption of the institute.

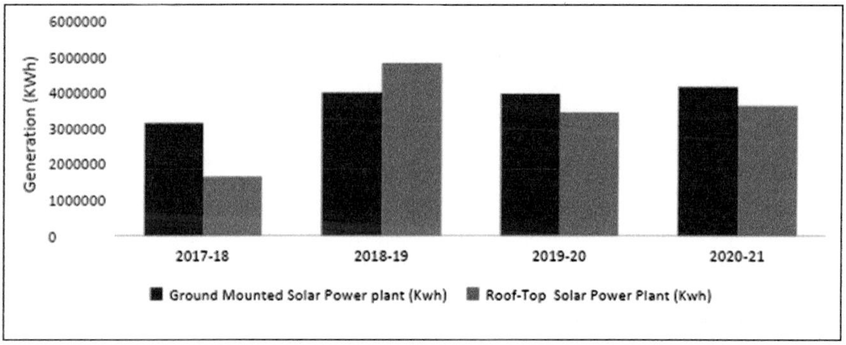

FIGURE 12.5 Electricity generation by a solar plant.

TABLE 12.1 Estimated Savings due to Solar Power Generation

Year	Total Cost	Per Unit Cost	Savings due to Solar	Growth Potential of Solar	Replacement Rate
2017–2018	28,744,259	6.0	21,471,408	0.17	0.83
2018–2019	43,825,588	5.0	54,544,165	0.30	0.70
2019–2020	37,178,758	5.0	42,828,383	0.27	0.73
2020–2021	39,514,517	5.1	43,492,786	0.37	0.63

12.6 CONCLUSION

Sustainable Development Goals (SDGs) are intertwined; therefore, collaboration across disciplines and industries is encouraged. Involvement of educational institutions is required in achieving the SDGs, particularly goals 7, 11, and 13. The SDGs deal with various social, economic, and environmental issues. In 2021, India is ranked 120th in the world in terms of attaining SDGs, which requires immediate action from all sectors of society. India has one of the largest networks of educational institutes. Educational institutions being an important component of a society have a responsibility to contribute to both local and global communities. Their involvement in achieving sustainable development is critical to promoting energy saving and optimization and developing an SDG-ready generation. They should be motivated to develop a strategy for achieving the SDGs. Such initiatives would help in reducing electricity expenses and fulfilling electricity demand for the achievement of SDGs. Educational institutes have already started coming forward in generating solar energy for running the educational institutes to save valuable resources for future generations. Roofs of educational institutions are being utilized for installing solar panels. Over the last decade, they have been doing great in achieving SDGs. The government is also bringing various schemes and programs to promote energy generation in all sectors through renewable sources to save the planet.

ACKNOWLEDGMENT

The author would like to thank to the Chairman, Department of Commerce, AMU, Aligarh, for providing facilities and a positive atmosphere. The author also thanks Prof. Shamim Ansari, Electricity Department, AMU and Dr. Salman Ansari, Electricity Department, AMU, for providing data, and assistance.

AUTHOR'S AGREEMENT

We declare that the work submitted for publication is original and previously unpublished in English or any other language(s) and not under consideration for publication elsewhere. We also certify that all the authors have approved the chapter for release and are in agreement with its content.

CREDIT AUTHORSHIP CONTRIBUTION STATEMENT

Jahangir Chauhan: Conceptualization, Methodology, Supervision. **Aqsa Anjum:** Data collection, Writing—review and drafting.

DECLARATION OF COMPETING INTEREST

The authors declare that they have no known competing financial interests or personal relationships that could have appeared to influence the work reported in this chapter.

KEYWORDS

- **academic institutions**
- **electricity cost**
- **energy saving**
- **solar energy**
- **solar panels**

REFERENCES

Ahmad, F., and Alam, M. S. (2017). Feasibility study, design and implementation of smart polygeneration microgrid at AMU. *Sustainable Cities and Society 35*, 309–322. https://doi.org/10.1016/j.scs.2017.08.007.

Ahmad, F., and Alam, M. S. (2017b). Optimal sizing and analysis of solar PV, wind, and energy storage hybrid system for campus microgrid. *Smart Science 6*(2), 150–157. https://doi.org/10.1080/23080477.2017.1417005.

Ahmad, F., and Iqbal, S. (2012). Reducing electricity consumption in educational institutes: A case study on Aligarh Muslim University's electricity usage scenario. In: *IEEE Students' Conference on Electrical, Electronics and Computer Science*. Bhopal: MANIT, pp. 34–38.

AMU Solar Power plant, Centre for Integrated Green and Renewable Energy, AMU. (Accessed March 21, 2022).

Ansari, M. S., Jalil, M., and Diwania, S. (2021). Techno-economic analysis of Diesel/Wind/PV/Battery hybrid energy system for Androth Island. *Renewable Power for Sustainable Growth* Singapore, pp. 285–295.

Baitule, A. S., and Sudhakar, K. (2017). Solar powered green campus: A simulation study. *International Journal of Low-Carbon Technologies 12*(4), 400–410.

Bano, T., and Rao, K. V. S. (2016). The effect of solar PV module price and capital cost on the levelized electricity cost of the solar PV power plant in the context of India. In: *2016 Biennial International Conference on Power and Energy Systems: Towards Sustainable Energy (PESTSE)*. IEEE, pp. 1–6.

Bhuyan, S., Singh, K., Kumar, M. N., and Mishra, S. (2017, December). A study on performance analysis of 75kWp grid connected solar power plant at MEMS. In: *2017 International Conference on Computer, Electrical & Communication Engineering (ICCECE)*. IEEE, pp. 1–5.

Khan, H., Siddiqui, A., Ansari, A. A., and Rafiuddin, N. (2021). Energy audit in an Indian University—A case study. In: *Renewable Power for Sustainable Growth*. Singapore: Springer, pp. 573–580.

Mendu, S. S., Appikonda, P., and Emadabathuni, A. K. (2020). Techno-economic comparative analysis between grid-connected and stand-alone integrated energy systems for an educational institute. *EVERGREEN Joint Journal of Novel Carbon Resource Sciences & Green Asia Strategy 7*(3), 382–395.

MERCOM. (2019). Aligarh Muslim University adds new solar rooftop; Total capacity reaches 6.5 MW—Mercom India. (Accessed February 27, 2019).

Nagar, R. R., and Gidwani, L. (2018). Levelized cost of electricity with degradation of 10 MW grid-connected photovoltaic power plant in Kalwakurthy, India. In: *2018 3rd International Innovative Applications of Computational Intelligence on Power, Energy and Controls with their Impact on Humanity (CIPECH)*. IEEE, pp. 50–55.

Ramana, P. V., Kumar, B. S., and Maheswar, D. (2015). A comprehensive study on solar power plant at educational institute level. *International Research Journal of Engineering and Technology 2*(04), 1063–1071.

Rupal, A., Syal, P., and Sharma, S. (2016). Energy conservation opportunities in institutional buildings-a case study in India. In: *2016 IEEE International Conference on Power and Renewable Energy (ICPRE)*. IEEE, pp. 664–668.

Tariq, M., Ansari, M. K., Rahman, F., Rahman, M. A., and Ashraf, I. (2021). Effect of soiling on the performance of solar PV modules: A case study of aligarh. *Smart Science 9*(2), 121–132. https://doi.org/10.1080/23080477.2021.1901340.

Werulkar, A. S., and Kulkarni, P. S. (2015). A case study of residential solar photovoltaic system with utility backup in Nagpur, India. *Renewable and Sustainable Energy Reviews 52*, 1809–1822. https://doi.org/10.1016/j.rser.2015.07.195.

CHAPTER 13

EXPLORING MARKET HIERARCHY AND INFRASTRUCTURAL FACILITIES FOR SUSTAINABLE RURAL DEVELOPMENT IN INDIA: A GEOGRAPHICAL PERSPECTIVE

MOHAMMAD MUQEET KHAN[1], TANVEER ASIF[2], and NIRAJ PRAKASH JOSHI[3,4]

[1]*Department of Geography, Jamia Millia Islamia University, New Delhi, India*

[2]*Department of Geography, Aligarh Muslim University, Aligarh, Uttar Pradesh, India*

[3]*Center for Peaceful and Sustainable Future (CEPEAS), Hiroshima University, Higashihiroshima, Hiroshima, Japan*

[4]*TAOYAKA Program, Hiroshima University, Higashihiroshima, Hiroshima, Japan*

ABSTRACT

This chapter aims to find out the hierarchy of selected rural markets as well as analyze the relationship between hierarchy of the markets and the development of infrastructure along with service facilities in north India. During the analysis, 20 chosen variables were examined to calculate the composite percent change in market infrastructure and service facilities from 1997

Sustainable Development: A Geospatial and Statistical Perspective. Qazi Mazhar Ali, Rizwan Ahmad, Irfan Ali, & Prasenjit Chatterjee (Eds.)

DOI: 10.1201/9781779643247-13

to 2017. The data have been collected at the market level through a field survey. The study highlighted that there is a direct relationship between the composite hierarchical score and the composite market facilities score of the selected markets. This relationship is more prevalent in block-level markets or medium-sized markets, which have shown considerable growth in market infrastructures, service facilities, and amenities.

13.1 INTRODUCTION

The market centers' hierarchical organization is a common occurrence. Although each market settlement has its unique status, it is also a crucial component of a regional system that interacts with other market settlements in both a functional and spatial way and is crucial for the sustainability of rural development (Saxena, 2004). The hierarchy notion, key to central place theory, has been widely utilized by geographers, economists, and scholars in various disciplines. In economic geography, it plays a vital role in determining the rankings of market centers and their spatial arrangement within a region. In general, hierarchy refers to the process of allocating ranks in sequential order, from lowest to highest, or *vice versa* (Khan et al., 2018). Market hubs can be rated and categorized into a variety of hierarchical class systems based on how they differ from one another. Each class offers specific items and commodities for trade while also serving as a foundation for sustainable rural development and planning.

The idea of hierarchy is roughly drawn from Christaller's classic work, Central Place Theory (1933), where he developed the idea of hexagonal market sectors based on the idea of a real range of commodities. Higher-order central places are said to be more widely spaced than lower-order central places, according to Christaller. Many academics have attempted to investigate the hierarchy of market centers in various ways. Skinner (1965) distinguished five tiers of markets: regional markets, minor markets, standard markets, central markets, and standardized markets. Berry (1967) determined seven degrees of hierarchy, namely hamlet, village, large village, small town, regional center, regional metropolis, and national metropolis based on many characteristics and their correlation. Berry's most well-known work in the USA is the classification of the hierarchy of central places in southwest Iowa, South Dakota, and Chicago.

Smailes (1944) used index functions to divide important locations in England and Wales into five hierarchical categories. Based on the functional hierarchy, Ukwu (1969) provided a thorough analysis of Ibo periodic marketplaces in eastern Nigeria. Using the functional composite indices, Ofori and Aseidu (2013) studied the hierarchy of market centers along the Volta Lake in Ghana. The composite index value has been used by several Indian geographers to look at the functional hierarchy of market centers in various regions. Some quantitative and qualitative attributes such as education, health, post office, banking, market size, and available infrastructural facilities have been used for calculating the Composite Functional Index. Similar methods to determine hierarchical order have been used by Vishwanath (1967), Jana (1978), Tamaskar (1981), Khan (1991), Mallick (2001), Dash (2005), and Khan et al. (2017).

13.2 TASKS AND STUDY AREA

The present study has been undertaken with the tasks of finding out the hierarchy of selected rural markets as well as analyzing the relationship between the hierarchy of selected markets and the development of infrastructure and service facilities. Further, to compute the composite percent change of 20 selected variables of change in market infrastructure and service facilities from 1997 to 2017 have been done. The variables have been obtained through the field survey at the market level during 2017.

The Aligarh district in the North Indian state of Uttar Pradesh has been selected as the study area (Figure 13.1). It is located in the northwestern part of Ganga Yamuna doab and forms a part of the Agra division and lies between 27° 27′ N to 28° 11′ N latitudes and 77° 27′ E to 78° 38′ E longitudes. The total area of the district is 3650 sq km, which supports a population of 3,673,849. Administratively, the district has been divided into five tehsils (subdivisions) and twelve development blocks. It has total of 1170 inhabited villages and 24 town areas.

13.3 DISCUSSION

13.3.1 COMPUTATION OF MARKET HIERARCHY

The study assessed the relative importance of market centers in the district and to group them into various hierarchical orders. The present hierarchical classification of selected rural markets of Aligarh district is based upon Composite Functional Index value computed from 11 parameters and 28 class of parameters including different quantitative and qualitative attributes (Table 13.1).

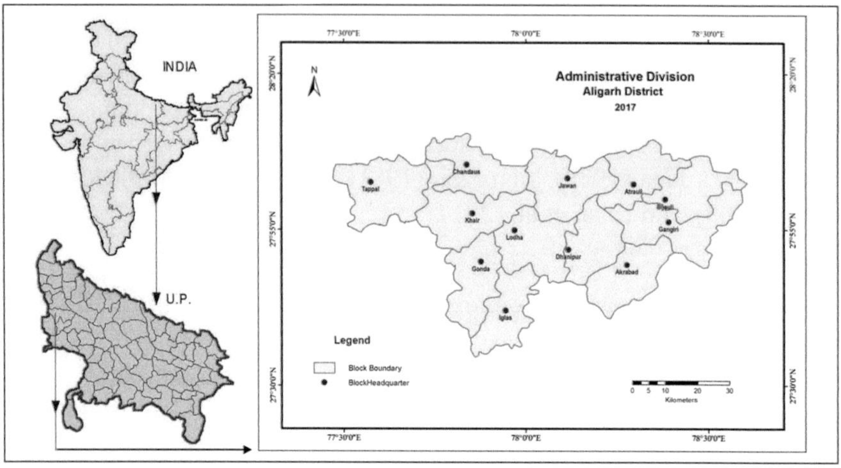

FIGURE 13.1 Map showing Aligarh district with its twelve development blocks.

Each attribute is given due weight keeping in view its importance for the particular rural market center. The subjective weighting technique has been applied to the Composite Functional Index. Institutional, infrastructural, and functional attributes are taken into account to understand the relative importance of market centers and to present them from a comprehensive perspective. The subjective principle is based on the researcher's understanding and judgment. However, strictly followed a 5-point weighting scale varying from 1 (lowest) to 5 (highest), applied logically following inter- and intraregional comparison and their relative importance for a market center. The computation of the Composite Functional Index is given in Appendix 13.1.

The Composite Functional Index (CFI) presented as Table 13.2 is computed using the given formula:

$$CFI = \sum_{i=1}^{n} FiWi$$

where,

CFI = Composite Functional Index
Fi = Number of markets/units
Wi = Weight of the particular function

TABLE 13.1 Weightage Scheme of Parameters

S. No.	Parameters	Class of Parameters	Weight
1	Administrative status	Tehsil Headquarters	3
2		Block Headquarters	2
3	Market area	50 sq km	1
4	Periodicity	Daily	3
5		Biweekly	2
6		Weekly	1
7	Participants	Nonpedestrian	2
8		Pedestrian	1
9	Market attendance	400 Participants	1
10	Commodity transaction	Large (>20 Commodities)	3
11		Medium (11–20 Commodities)	2
12		Small (≤ 10 Commodities)	1
13	Shops	50 Shops in a Market	1
14	Distance	Average 1 km distance travelled by traders	1
15		Average 1 km distance travelled by consumers	1
16	Public amenities	Schools and colleges	5
17		Police station	4
18		Post office	3
19		Petrol pump	2
20		Marriage hall	1
21	Transportation	Railway station or halt	3
22		Bus stand	2
23		Taxi and auto	1
24	Connectivity and accessibility	National highway	5
25		State highway	4
26		District road	3
27		Metalled road	2
28		Unmetalled road	1

TABLE 13.2 CFI Value of Selected Rural Markets in Aligarh District

S. No.	Selected Markets	CFI Value
1	Roravar	252.04
2	Pisava	172.47
3	Nagla Sabal *urf* Gonda	160.28
4	Tappal	155.54
5	Chandaus	147.68
6	Jirauli Dhoom Singh	120.47
7	Akrabad	115.38
8	Narauna Akapur	99.83
9	Kazimabad	88.12
10	Jawan Sikanderpur	82.49
11	Gopi	54.96
12	Gomat	54.06
13	Panaithi	52.42
14	Barla	49.52
15	Hastpur Chandfari	39.72
16	Nagla Darvar	37.94
17	Lodha	36.43
18	Bijauli	34.45
19	Shivala	29.36
20	Barhaul	24.26
21	Sathini	22.11
22	Narupura Katka	20.89
23	Jartauli	18.37
24	Rajmau	17.77
25	Dhansari	16.88
26	Goraula	14.51
27	Sikandarpur Bhukarawale	14.48
28	Kalai	11.59
29	Bhanauli	10.94
30	Sunamai	10.43

Source: Computed by the researcher on the basis of field survey.

Thus, on the basis of calculated Composite Functional Index, the rural markets are classified into three grades; namely regional market (>100 CFI), block level market (50–100 CFI), and local market (< 50 CFI) (Table 13.3).

TABLE 13.3 Hierarchy of Selected Rural Markets in Aligarh District

Order	Category	Name of Market	No. of Markets	Composite Functional Index
I	Regional Market	Roravar, Pisava, Nagla Sabal *urf* Gonda, Tappal, Chandaus, Jirauli Dhoom Singh, Akrabad	7	> 100.00
II	Block Level Market	Narauna Akapur, Kazimabad, Jawan Sikanderpur, Gopi, Gomat, Panaithi	6	50.00–100.00
III	Local Market	Barla, Hastpur Chandfari, Nagla Darvar, Lodha, Bijauli, Shivala, Barhaul, Sathini, Narupura Katka, Jartauli, Rajmau, Dhansari, Goraula, Sikandarpur Bhukarawale, Kalai, Bhanauli, Sunamai	17	< 50.00

The study found three orders of market hierarchy in the study area. There are 7 first-order markets (Regional markets) namely, *Roravar, Pisava, Nagla Sabal urf Gonda, Tappal, Chandaus, Jirauli Dhoom Singh, Akrabad*, 6 second-order markets (Block level markets), that is, *Narauna Akapur, Kazimabad, Jawan Sikanderpur, Gopi, Gomat, Panaithi*, and 17 third-order markets (Local markets), namely, *Barla, Hastpur Chandfari, Nagla Darvar, Lodha, Bijauli, Shivala, Barhaul, Sathini, Narupura Katka, Jartauli, Rajmau, Dhansari, Goraula, Sikandarpur Bhukarawale, Kalai, Bhanauli, Sunamai* (Table 13.3 and Figure 13.2).

13.3.2 COMPOSITE HIERARCHICAL SCORE VIS-À-VIS COMPOSITE MARKET-FACILITIES SCORE

The role of market centers in the development of infrastructure and service facilities is quite clear from the ongoing discussions. For the study, a *Composite Hierarchical Score (CHS)-Composite Market-Facilities Score* (CMS) M*odel* has been developed. The various characteristics pertaining to a market when collectively scored form the CHS. Thus, the CHS is an indicator of the hierarchy of a particular market among the 30 selected markets. For the present study, *28 quantitative and qualitative attributes* available in selected rural markets were analyzed (Appendix 13.1). Table 13.4 shows the hierarchical classification of 30 selected markets on the basis of their individual CHS. Likewise, the *20 variables* pertaining to various market infrastructures and service facilities (Appendix 13.2) of the selected markets collectively represent the CMS.

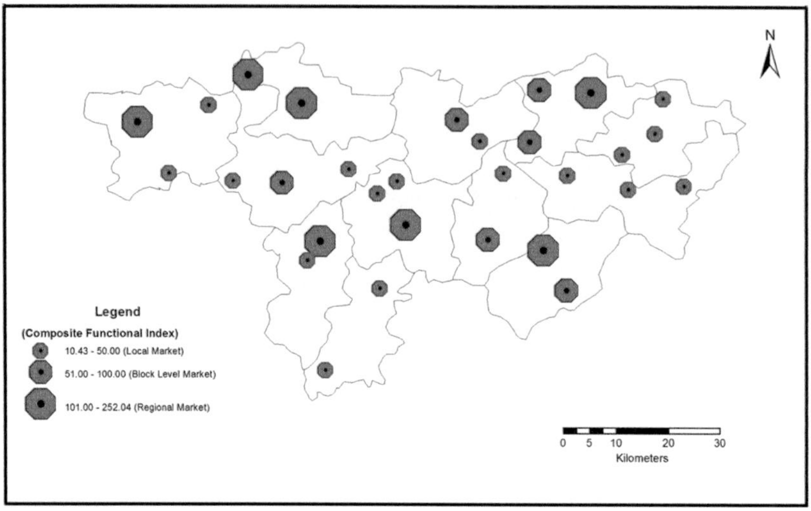

FIGURE 13.2 Hierarchy of selected rural markets.

Thus, the CMS of a particular selected market is an indicator of the rate of change of market infrastructure and service facilities of that particular market among the 30 selected markets. Table 13.4 also shows the Composite Market-facilities Score of the selected rural markets in the Aligarh district.

Figures 13.3 and 13.4 show the relationship between a particular market's hierarchy and its relation to the development of infrastructure and service facilities. The study highlighted that the comparative trend of the CHS and CMS of the selected markets shows that there is an overall increase in CMS with the increasing CHS.

Thus, it reveals that the market of higher order, that is, of higher CHS is also having a higher CMS. The above results indicate that the role of a particular rural market in the development of infrastructure and service facilities is highly affected by its hierarchy. Although, some markets do not follow the general trend and have shown greater CMS with respect to their hierarchy among the selected markets. Thus, these markets have a greater role in the development of infrastructure and service facilities than the markets of a higher hierarchy. Thus, *Narauna Akapur, Kazimabad, Jawan Sikanderpur, Gopi, Gomat* are markets that fall in this category (Figure 13.3).

TABLE 13.4 Composite Hierarchical Score (CHS) *vis-à-vis* Composite Market-facilities Score (CMS) of Selected Rural Markets in Aligarh District

S. No.	Selected Market	CHS	CMS
Regional Markets			
1	Roravar	7561.49	8799.90
2	Pisava	5174.30	8952.60
3	Nagla Sabal *urf* Gonda	4808.42	10,524.94
4	Tappal	4666.35	9455.10
5	Chandaus	4430.60	10,139.30
6	Jirauli Dhoom Singh	3614.25	8516.93
7	Akrabad	3461.43	8445.68
Block level markets:			
8	Narauna Akapur	3054.96	11,636.37
9	Kazimabad	2643.70	10,780.52
10	Jawan Sikanderpur	2474.94	9477.81
11	Gopi	1648.91	8764.66
12	Gomat	1621.81	10,077.48
13	Panaithi	1572.72	7355.58
Local Markets			
14	Barla	1485.84	7258.69
15	Hastpur Chandfari	1191.63	8074.76
16	Nagla Darvar	1138.48	8233.60
17	Lodha	1093.10	6403.33
18	Bijauli	1033.69	7269.84
19	Shivala	881.00	5682.80
20	Barhaul	727.88	5453.31
21	Sathini	663.46	5034.92
22	Narupura Katka	626.72	3451.56
23	Jartauli	551.18	4160.59
24	Rajmau	533.18	4727.54
25	Dhansari	506.59	4142.97
26	Goraula	435.50	3311.90
27	Sikandarpur Bhukarawale	434.57	2973.68
28	Kalai	347.75	3832.65
29	Bhanauli	328.44	2980.76
30	Sunamai	312.93	3624.32

Source: Computed by the researcher on the basis of field survey.

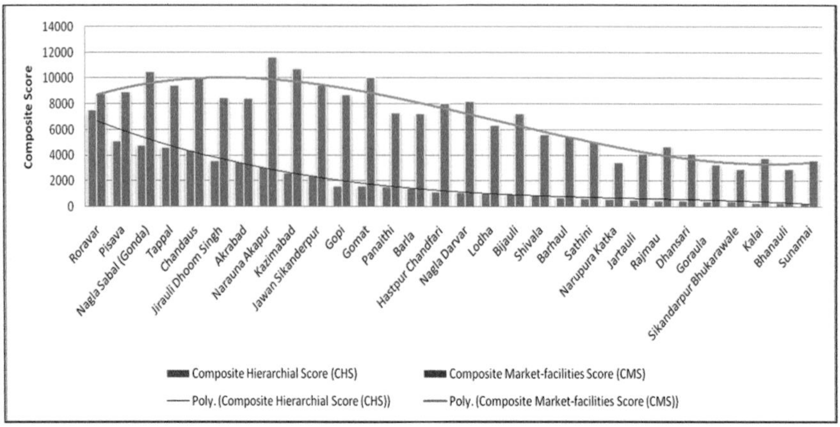

FIGURE 13.3 Relationship between CHS and CMS.

The role of market hierarchy in the development of infrastructure and service facilities is further evident in Figure 13.4. It reveals that there is a direct relationship between the CHS and CMS of the selected markets, that is, among the rate of development of infrastructure and service facilities and the hierarchy of selected rural markets. This shows that although each market has developed infrastructure and service facilities in the study area, the rate of development of infrastructure and service facilities has been found highest in the case of block-level markets or medium-sized markets. This is due to the fact that these markets have shown considerable growth in market infrastructure, service facilities, and amenities with respect to the large-sized markets.

13.4 CONCLUSION

In view of the importance of the market order in the development of infrastructure in the study region, the present study aims to find out the hierarchy of selected rural markets as well as analyze the relationship between market hierarchy and the development of infrastructure along with service facilities. The study found three orders of market hierarchy, there are 7 first-order markets (regional markets), 6 second-order markets (block-level markets), and 17 third-order markets (local markets).

The *CHS-CMS Model* developed in the study highlighted that there is an overall increase in CMS with the increasing CHS, that is, infrastructure increases with an increase in the hierarchy. The study concluded that the medium-sized markets have a higher rate of development of infrastructure and service facilities than the large-sized rural markets due to their considerable growth during this period. These markets have a significant role in developing infrastructural facilities and services in their hinterland and generating employment opportunities for the sustainable development of the rural areas. Thus, this market order needs to be focused on for further development and should be included in the area development plans.

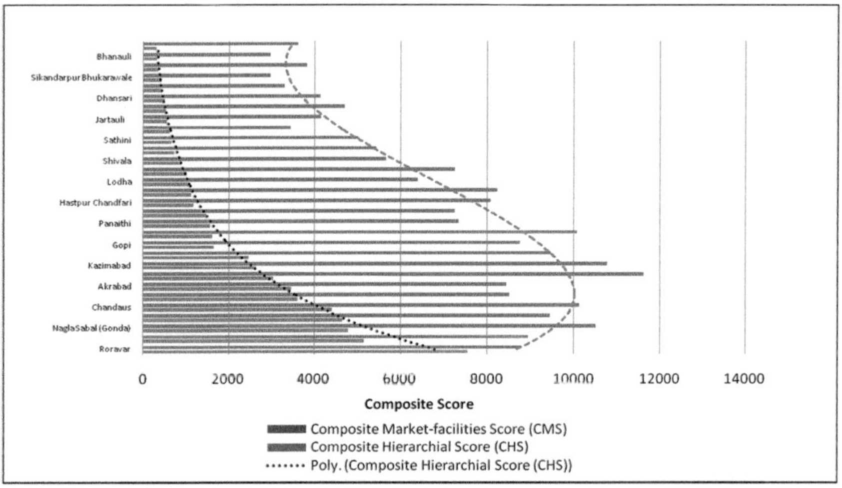

FIGURE 13.4 The comparative trend of CHS and CMS in selected market centers.

KEYWORDS

- **markets**
- **hierarchy**
- **infrastructural facilities**
- **sustainable rural development**

REFERENCES

Berry, B. J. L. (1967). *Geography of Market Centers and Retail Distribution*. New Jersey.

Christaller, W. (1933). *Central Places in Southern Germany*. Baskin, C. W. (Ed.). New Jersey: Prentice Hall, Englewood Cliffs, 1966.

Dash, N. (2005). Hierarchy of market centers in the Rural Hinterland of Cuttack-Bhubaneswar in Orissa, India. *Indian Journal of Regional Science 37*(1), 120–130.

Jana, M. M. (1978). Hierarchy of market centers in lower silabati basin. *Geographical Review of India 40*, 164–174.

Khan, M. M., Khan, N., and Asif, T. (2017). Hierarchy of market centers and rural development: Evidences from North India. *Research Forum: International Journal of Social Sciences 5*(1), 20–28.

Khan, M. M., Khan, N., and Asif, T. (2018). *Rural Markets and Rural Transformation: The Changing Rural Realities*. New Delhi: R. P. Publications.

Khan, N. (1991). *Agricultural Development and Marketing*. Delhi: H.K. Publishers.

Mallick, R. K. (2001). Identification and accessibility analysis of rural service centers in Kendrapa District, Orrisa, India: A GIS based application. *International Journal of Applied Earth Observation and Geo-informatics 3*, 99–106.

Ofori, B. D., and Aseidu, A. B. (2013). Hierarchy of market centers along the Volta Lake in Ghana: A regional development framework. *Journal of Geography and Regional Planning 6*(8), 280–293.

Saxena, H. M. (2004). *Marketing Geography*. New Delhi: Rawat Publication.

Skinner, G. W. (1965). Marketing and social structure in Rural China. *Journal of Asian Studies 24*, 3–45.

Smailes, A. E. (1944). Urban hierarchy in England and Wales. *Geography 29*, 41–51.

Tamaskar, B. G. (1981). Market attractiveness of arvi upland. *Geographical Review of India 43*, 4.

Ukwu, U. I. (1969). Markets in Iboland. In: Hodder, B. W., Ukwu, U. I. (Eds.). *Markets in West Africa*, Ibadan.

Vishwanath, M. S. (1967). *A Geographical Analysis of Rural Markets and Urban Centers in Mysore*. Unpublished Ph.D. Thesis, Varanasi: Department of Geography, Banaras Hindu University.

APPENDIX 13.1 Computation of Composite Functional Index to Identify the Hierarchy of Selected Rural Markets in Aligarh District

S. No.	Selected Market	1		2		3		4		5		6		7		8		9	
		Administrative HQ	**Score T. HQ=3 B. HQ=2**	**Market Trade Area**	**Score 50 Sq Km=1**	**Periodicity**	**Score D=3 Bw=2 W=1**	**Pedestrian**	**Score One person=1**	**Non-Pedestrian**	**One person=2**	**Market Attendance**	**Score 400 parts=1**	**Commodities**	**Score L=3 M=2 S=1**	**Shops**	**Score 50 shops=1**	**Trader Distance**	**Score 1 km=1**
1	Tappal	B.HQ	2	173.34	3.5	D,W	4	1000	1000	1800	3600	2800	7	L	3	500	10	4.96	4.96
2	Jartauli	–	–	17.64	0.25	D,W	4	225	225	150	300	375	0.85	S	1	70	1.35	2.61	2.61
3	Goraula	–	–	15.90	0.2	D,W	4	200	200	100	200	300	0.75	S	1	50	1	2.62	2.62
4	Chandaus	B.HQ	2	203.48	4	D,Bw	5	800	800	1780	3560	2580	6.5	L	3	450	9	5.96	5.96
5	Pisava	–	–	155.18	3.5	D,Bw	5	1110	1110	2000	4000	3110	7.75	L	3	550	11	5.59	5.59
6	Gomat	–	–	36.51	0.5	D	3	380	380	600	1200	980	2.5	M	2	200	4	2.44	2.44
7	Shivala	–	–	20.90	0.35	D,W	4	250	250	300	600	550	1.25	S	1	110	2.25	2.72	2.72
8	Bhanauli	–	–	18.85	0.25	D,W	4	80	80	110	220	190	0.5	S	1	40	0.80	2.83	2.83
9	Jawan Sikanderpur	B.HQ	2	69.95	0.75	D,W	5	620	620	900	1800	1520	3.75	L	3	300	6	3.74	3.74
10	Sunamai	–	–	20.26	0.35	D,W	5	120	120	80	160	200	0.5	S	1	50	1	2.52	2.52
11	Roravar	–	–	99.18	2	D	3	1500	1500	3000	6000	4500	11.25	L	3	800	16	3.74	3.74
12	Lodha	B.HQ	2	29.21	0.5	D	3	250	250	400	800	650	1.5	M	2	150	3	2.86	2.86
13	Sikandarpur Bhukarawale	–	–	17.19	0.25	D,W	4	100	100	150	300	250	0.75	S	1	45	0.9	1.94	1.94
14	Panaithi	–	–	43.92	0.5	D	3	400	400	560	1120	960	2.5	L	3	175	3.25	3.59	3.59
15	Kalai	–	–	15.90	0.25	W	1	175	175	75	150	250	0.75	S	1	65	1.25	2.64	2.64
16	Nagla Sabal *urf* Gonda	B.HQ	2	193.00	3.25	D,Bw	5	1150	1150	1800	3600	2950	2.5	L	3	500	10	5.78	5.78

APPENDIX 13.1 *(Continued)*

S. No.	Selected Market	1		2		3		4		5		6		7		8		9	
		Administrative HQ	Score T. HQ=3 B. HQ=2	Market Trade Area	Score 50 Sq Km=1	Periodicity	Score D=3 Bw=2 W=1	Pedestrian	Score One person=1	Non-Pedestrian	One person=2	Market Attendance	Score 400 parts=1	Commodities	Score L=3 M=2 S=1	Shops	Score 50 shops=1	Trader Distance	Score 1 km=1
17	Nagla Darvar	-	-	70.55	1.5	D,Bw	5	300	300	400	800	700	1.75	M	2	135	2.75	3.74	3.74
18	Hastpur Chandfari	-	-	30.96	0.5	D	3	350	350	40	800	750	1.85	M	2	150	3	2.77	2.77
19	Sathini	-	-	20.90	0.25	D	3	250	250	190	380	440	1.15	M	2	90	1.9	2.43	2.43
20	Kazimabad	-	-	56.18	1.25	D	3	600	600	1000	2000	1600	4	L	3	400	8	3.08	3.08
21	Jirauli Dhoom Singh	-	-	140.11	2.75	D	3	950	950	1300	2600	2,250	5.65	L	3	425	8.5	4.03	4.03
22	Narauna Akapur	-	-	142.22	2.75	D,Bw	5	800	800	1100	2200	1900	4.75	M	2	350	7	5.60	5.60
23	Bijauli	B.HQ	2	28.64	0.5	D,Bw	5	300	300	350	700	650	1.5	S	1	170	3.65	3.23	3.23
24	Barhaul	-	-	21.88	0.35	D,Bw	5	260	260	220	440	480	1.25	M	2	100	2	2.45	2.45
25	Narupura Katka	-	-	18.69	0.25	D,W	4	160	160	220	440	380	0.85	S	1	80	1.75	2.33	2.33
26	Dhansari	-	-	16.61	0.25	D	3	125	125	175	350	300	0.75	S	1	50	1	2.17	2.17
27	Rajmau	-	-	18.69	0.25	W	1	210	210	150	300	360	0.80	S	1	75	1.25	2.44	2.44
28	Barla	-	-	35.66	0.75	D,W	4	450	450	500	1000	950	2.35	M	2	200	4	3.00	3.00
29	Akrabad	B.HQ	2	85.56	1.75	D,W	4	800	800	1300	2600	2100	5.25	L	3	450	9	3.95	3.95
30	Gopi	-	-	46.06	0.75	D	3	400	400	600	1200	1000	2.5	L	3	300	6	2.70	2.70

APPENDIX 13.1 *(Continued)*

10		11		12		13			
Consumer Distance	**Score 1 km=1**	**Public Amenities**	**Score SC=5 PS=4 PO=3 PP=2 MH=1**	**Public Transport**	**Score R/H=3 BS=2 T/A=1**	**Connectivity**	**Score NH=5 SH=4 DR=3 MR=2 UmR=1**	**Composite Score**	**Composite Functional Index (CFI) $\sum fx/n$**
9.89	9.89	SC,PO,PP	10	BS	2	SH,DR,MR,UmR	10	**4666.35**	**155.54**
2.12	2.12	SC,PO	8	–	–	DR,MR,UmR	6	**551.18**	**18.37**
1.88	1.88	SC,PO	8	T/A	1	NH,SH,DR,MR,UmR	15	**435.50**	**14.51**
10.14	10.14	SC,PS,PO	12	BS,T/A	3	SH,DR,MR,UmR	10	**4430.60**	**147.68**
8.46	8.46	SC,PO,PP,MH	11	BS,T/A	3	DR,MR,UmR	6	**5174.30**	**172.47**
4.37	4.37	SC,PO,PP	10	BS,T/A	3	SH,DR,MR,UmR	10	**1621.81**	**54.06**
2.43	2.43	SC,PO	8	BS,T/A	3	DR,MR,UmR	6	**881.00**	**29.36**
2.06	2.06	SC,PO	8	R/H,BS,T/A	6	DR	3	**328.44**	**10.94**
5.70	5.70	SC,PS,PO	12	BS,T/A	3	SH,DR,MR,UmR	10	**2474.94**	**82.49**
2.56	2.56	SC,PO	8	R/H,BS,T/A	6	DR,MR,UmR	6	**312.93**	**10.43**
7.50	7.50	SC,MH	6	BS,T/A	3	DR,MR,UmR	6	**7561.49**	**252.04**
3.24	3.24	SC,PS,PO	12	BS,T/A	3	SH,DR,MR,UmR	10	**1093.10**	**36.43**
2.73	2.73	SC	5	BS,T/A	3	NH,SH,DR,MR,UmR	15	**434.57**	**14.48**
3.88	3.88	SC,PS,PO,PP,MH	15	BS,T/A	3	NH,SH,DR,MR,UmR	15	**1572.72**	**52.42**
1.86	1.86	SC,PO	8	-	-	DR,MR,UmR	6	**347.75**	**11.59**
9.89	9.89	SC,PO	8	BS,T/A	3	DR,MR,UmR	6	**4808.42**	**160.28**
5.74	5.74	SC,PO	8	BS	2	DR,MR,UmR	6	**1138.48**	**37.94**
3.51	3.51	SC,PS,PO	12	BS,T/A	3	SH,DR,MR,UmR	10	**1191.63**	**39.72**

APPENDIX 13.1 *(Continued)*

10		11		12		13			
Consumer Distance	Score 1 km=1	Public Amenities	Score SC=5 PS=4 PO=3 PP=2 MH=1	Public Transport	Score R/H=3 BS=2 T/A=1	Connectivity	Score NH=5 SH=4 DR=3 MR=2 UmR=1	Composite Score	Composite Functional Index (CFI) ∑fx/n
2.73	2.73	SC,PO	8	BS	2	SH,DR,MR,UmR	10	**663.46**	**22.11**
5.37	5.37	SC,PO	8	BS	2	DR,MR,UmR	6	**2643.70**	**88.12**
9.32	9.32	SC,PS,PO,PP,MH	15	BS,T/A	3	SH,DR,MR,UmR	10	**3614.25**	**120.47**
7.86	7.86	SC,PO	8	BS	2	SH,DR,MR,UmR	10	**3054.96**	**99.83**
2.81	2.81	PO,PP	5	BS,T/A	3	DR,MR,UmR	6	**1033.69**	**34.45**
2.83	2.83	SC,PP	7	-	-	DR,MR	5	**727.88**	**24.26**
2.54	2.54	SC,PO	8	-	-	DR,MR,UmR	6	**626.72**	**20.89**
2.42	2.42	SC,PO,MH	9	BS	2	SH,DR,MR,UmR	10	**506.59**	**16.88**
2.44	2.44	SC,PO	8	-	-	DR,MR,UmR	6	**533.18**	**17.77**
3.74	3.74	SC,PO	8	BS	2	DR,MR,UmR	6	**1485.84**	**49.52**
6.48	6.48	SC,PO	8	BS,T/A	3	NH,SH,DR,MR,UmR	15	**3461.43**	**115.38**
4.96	4.96	SC,PO	8	BS,T/A	3	NH,SH,DR,MR,UmR	15	**1648.91**	54.96

Source: Computed by the researcher on the basis of field survey.

T. HQ = Tahsil Headquarters, B. HQ = Block Headquarters, D = Daily market, Bw = Bi-weekly market, W = Weekly market, L = Large no. of commodities (15-22), M = Medium no. of commodities (9-14), S = Small no. of commodities (5-7), B = Bank, HS = Health Service Center, PO = Post Office, BS = Bus Stand, TS = Taxi Stand, NH = National Highway, SH = State Highway, MR = Metalled Road, J = Railway Junction/ Halt, UmR = Unmetalled Road, SC = Schools & Colleges, PS = Police Station, PO = Post Office, PP = Petrol Pump, MH = Marriage Hall.

APPENDIX 13.2 Composite Percent Change (1997–2017) of Selected Variables of Change in Market Infrastructure and Service Facilities in Selected Rural Markets of Aligarh District

1	2	3	4	5	6	7	8	9	10	11	12	13	14	15	16	17	18	19	20	21	22
Selected Market	**X1**	**X2**	**X3**	**X4**	**X5**	**X6**	**X7**	**X8**	**X9**	**X10**	**X11**	**X12**	**X13**	**X14**	**X15**	**X16**	**X17**	**X18**	**X19**	**X20**	**Composite**
Tappal	150.00	200.00	650.00	500.00	400.00	1500.00	100.00	100.00	466.66	462.50	500.00	766.66	675.00	500.00	500.00	733.33	600.00	300.00	266.66	84.29	**9455.10**
Jartauli	0	200.00	200.00	0	0	300.00	0	0	525.00	700.00	300.00	500.00	200.00	0	300.00	300.00	400.00	0	200.00	35.59	**4160.59**
Goraula	100.00	300.00	300.00	0	0	200.00	0	0	233.33	550.00	200.00	300.00	200.00	0	200.00	200.00	300.00	0	200.00	28.57	**3311.90**
Chandaus	100.00	300.00	433.33	150.00	300.00	650.00	100.00	100.00	420.00	320.00	1000	1300.00	800.00	400.00	933.33	1050.00	900.00	400.00	400.00	82.64	**10,139.30**
Pisava	200.00	600.00	275.00	200.00	300.00	750.00	100.00	0	214.28	300.00	766.66	1100.00	700.00	550.00	450.00	833.33	533.33	300.00	700.00	80.00	**8952.60**
Gomat	200.00	400.00	700.00	100.00	100.00	500.00	100.00	0	400.00	200.00	700.00	900.00	1500.00	500.00	1400.00	1000.00	800.00	0	500.00	77.48	10,077.48
Shivala	100.00	400.00	500.00	100.00	0	200.00	100.00	0	1233.33	200.00	200.00	400.00	500.00	0	500.00	500.00	400.00	0	300.00	49.47	**5682.80**
Bhanauli	100.00	200.00	200.00	100.00	0	200.00	0	0	450.00	300.00	200.00	200.00	200.00	0	200.00	200.00	200.00	0	200.00	30.76	**2980.76**
Jawan Sikanderpur	200.00	250.00	550.00	100.00	200.00	1,100.00	100.00	100.00	450.00	240.00	400.00	650.00	800.00	500.00	950.00	1500.00	500.00	200.00	600.00	87.81	**9477.81**
Sunamai	0	200.00	500.00	0	0	400.00	0	0	550.00	250.00	200.00	200.00	300.00	0	200.00	200.00	200.00	0	400.00	24.32	**3624.32**
Roravar	200.00	200.00	600.00	100.00	200.00	566.66	0	100.00	400.00	300.00	600.00	650.00	660.00	650.00	675.00	1233.33	525.00	500.00	550.00	89.91	**8799.90**
Lodha	200.00	300.00	233.33	100.00	100.00	400.00	100.00	0	300.00	200.00	300.00	300.00	700.00	300.00	1,000.00	700.00	300.00	200.00	600.00	70.00	**6403.33**
Sikandarpur Bhukarawale	0	200.00	400.00	0	0	200.00	0	0	500.00	400.00	100.00	100.00	200.00	0	200.00	200.00	200.00	0	200.00	73.68	**2973.68**
Panaithi	200.00	200.00	300.00	100.00	100.00	700.00	100.00	100.00	286.66	212.50	66.66	300.00	1000.00	300.00	1200.00	900.00	600.00	0	600.00	89.76	**7355.58**
Kalai	0	0	300.00	100.00	0	200.00	0	0	700.00	200.00	400.00	600.00	300.00	0	300.00	300.00	200.00	0	200.00	32.65	**3832.65**
Nagla Sabal *urf* Gonda	250.00	350.00	433.33	400.00	300.00	700.00	100.00	100.00	427.77	283.33	1150.00	833.33	725.00	550.00	1066.66	1200.00	666.66	500.00	400.00	88.86	10,524.94
Nagla Darvar	100.00	200.00	650.00	100.00	100.00	900.00	0	0	380.00	414.28	200.00	400.00	800.00	200.00	900.00	800.00	1000.00	400.00	600.00	89.32	**8233.60**
Hastpur Chandfari	200.00	400.00	450.00	100.00	100.00	900.00	100.00	0	350.00	460.00	250.00	400.00	900.00	300.00	1000.00	700.00	800.00	300.00	300.00	64.76	**8074.76**
Sathini	0	400.00	400.00	100.00	0	400.00	0	0	600.00	600.00	300.00	300.00	400.00	0	600.00	400.00	200.00	0	300.00	34.92	**5034.92**
Kazimabad	300.00	300.00	233.33	200.00	100.00	400.00	100.00	0	511.11	487.50	566.66	700.00	1700.00	400.00	2500.00	900.00	900.00	0	400.00	81.92	**10,780.52**

APPENDIX 13.2 *(Continued)*

1	2	3	4	5	6	7	8	9	10	11	12	13	14	15	16	17	18	19	20	21	22
Selected Market	**X1**	**X2**	**X3**	**X4**	**X5**	**X6**	**X7**	**X8**	**X9**	**X10**	**X11**	**X12**	**X13**	**X14**	**X15**	**X16**	**X17**	**X18**	**X19**	**X20**	**Composite**
Jirauli Dhoom Singh	150.00	200.00	600.00	100.00	200.00	1100.00	100.00	0	290.47	320.00	350.00	650.00	633.33	450.00	1350.00	633.33	500.00	400.00	400.00	89.80	**8516.93**
Narauna Akapur	150.00	250.00	600.00	100.00	200.00	1400.00	100.00	0	885.71	416.66	700.00	1200.00	2000.00	250.00	1100.00	750.00	650.00	300.00	500.00	84.00	**11,636.37**
Bijauli	100.00	400.00	700.00	200.00	100.00	900.00	100.00	0	428.57	566.66	400.00	300.00	600.00	200.00	400.00	800.00	300.00	200.00	500.00	74.61	**7269.84**
Barhaul	200.00	300.00	800.00	0	0	300.00	0	0	900.00	275.00	200.00	200.00	300.00	200.00	700.00	500.00	200.00	0	300.00	78.31	**5453.31**
Narupura Katka	100.00	200.00	200.00	0	0	200.00	0	0	550.00	350.00	300.00	300.00	200.00	100.00	200.00	300.00	200.00	0	200.00	51.56	**3451.56**
Dhansari	200.00	300.00	500.00	0	0	400.00	0	0	300.00	266.66	200.00	300.00	300.00	300.00	200.00	200.00	200.00	0	400.00	76.31	4142.97
Rajmau	100.00	0	300.00	0	0	300.00	0	0	1000.00	275.00	400.00	400.00	400.00	200.00	300.00	300.00	300.00	0	400.00	52.54	**4727.54**
Barla	300.00	300.00	500.00	100.00	100.00	400.00	100.00	0	372.72	271.42	500.00	900.00	350.00	400.00	1200.00	400.00	200.00	300.00	500.00	64.55	**7258.69**
Akrabad	300.00	200.00	300.00	300.00	300.00	700.00	100.00	0	537.50	488.88	450.00	850.00	733.33	300.00	933.33	633.33	333.33	400.00	500.00	85.98	**8445.68**
Gopi	200.00	400.00	450.00	100.00	200.00	900.00	0	0	592.30	487.50	400.00	1100.00	700.00	250.00	1400.00	400.00	400.00	300.00	400.00	84.86	8764.66

Source: Computed by the researcher on the basis of data collected through field survey.

Selected Variables of Change in Market Infrastructure and Service Facilities	
X_1	Percentage of Godowns
X_2	Percentage of Agri-Clinics/Agri-Business Centers
X_3	Percentage of Agricultural Implements and Fertilizer Shops
X_4	Percentage of Banks
X_5	Percentage of ATMs
X_6	Percentage of Medical Centers
X_7	Percentage of Veterinary Hospitals
X_8	Percentage of Gas Agencies
X_9	Percentage of Grocery Stores
X_{10}	Percentage of Meat, Vegetable, Fruit and Dairy Product Shops
X_{11}	Percentage of Electrical Shops
X_{12}	Percentage of Electronics Shops
X_{13}	Percentage of Public Communication Centers (PCO/Mobile Shops)
X_{14}	Percentage of DJ Band/Musical Instrument Centers
X_{15}	Percentage of Building Materials and Sanitary ware Shops
X_{16}	Percentage of Garment Shops
X_{17}	Percentage of Stationery Shops
X_{18}	Percentage of Internet Cafes
X_{19}	Percentage of Bicycle, Bike, and Automobile Agencies/Service Centers
X_{20}	Percentage Share of Respondents Acknowledging Increase in Public Transport Infrastructure

CHAPTER 14

QUANTIFICATION OF POTENTIAL EVAPOTRANSPIRATION FROM PINDARI–KAFNI GLACIER VALLEYS IN KUMAUN CENTRAL HIMALAYAS

PANKAJ CHAUHAN, SAMEER TIWARI, and KALACHAND SAIN

Wadia Institute of Himalayan Geology, Dehradun, India

ABSTRACT

This chapter deals with the quantification of potential evapotranspiration (ETo) from glacier valleys of the Central Himalayas, namely, Pindar and Kafni. The Pindari and Kafni glaciers lie in the Pindar valley and drain into the Alaknanda River (upper Ganga basin) in the Bagehswar district of Uttarakhand state. In the study, ETo was computed with the *in situ* meteorological data using the Penman–Monteith equation, with two standard methods: (1) FAO method and (2) Cropwat 8.0. Results achieved from both the methods were compared and validated from modeled Global Land Cover Facility Modis data. Three-year (2017–2019) ETo based on evapotranspiration was quantified. The annual ETo observed in the whole Pindari valley was 2.49 mm day^{-1} (SD ± 0.39). Furthermore, the results were presented season wise; the results show that maximum ETo was computed (3.41 mm day^{-1} ± 0.07) in the summer season and minimum ETo was observed (1.35 ± 0.30 mm day^{-1}) in the winter season. Spring and autumn season ETo was observed (3.04 ± 0.55 mm day^{-1}) and (2.17 ± 0.65 mm day^{-1}), respectively. Results show variability from different methods and observations, which resulted that the patterns of the

Sustainable Development: A Geospatial and Statistical Perspective. Qazi Mazhar Ali, Rizwan Ahmad, Irfan Ali, & Prasenjit Chatterjee (Eds.)

DOI: 10.1201/9781779643247-14

ETo from both applied methods and modeled Modis data are the same in months in three years (2017–2019) average. Studied period monthly ETo was observed in the months of June and July. Observed ET was validated with MODIS reanalysis data and achieved the value $R^2 = 0.70$ and 0.56 FAO and Cropwat, respectively, which shows good correlation. Season-wise maximum ETo was calculated in summer and minimum in winter. Humidity and rainfall are the key factors to loss water in the valley in the context of the hydrological cycle.

14.1 INTRODUCTION

In recent years, extensive research was conducted vis-à-vis the trends in evapotranspiration (ET) under the potential impact of climate change in various regions over the globe at different time periods with different types of climate (e.g., Yin et al., 2010a, 2010b; Dinpashoh et al., 2011; Papaioannou et al., 2011; Kousari and Ahani, 2012; Zhao et al., 2015; Liu et al., 2016; Khanmohammadi et al., 2018, Yao et al., 2018).

Potential evapotranspiration (PET) could potentially use to estimate the atmospheric water demand of a specific area or region, which helps in numerous ways such as planning irrigation systems, tracking droughts, and analyzing the implications of climate change. Additionally, in recent years, several different models based on remote sensing have employed PET, or reference ET, to calculate actual ET (Allen et al., 2007; Senay et al., 2007).

Numerous methods have been testified for estimating the PET; nevertheless, it is tough to choose the best fit for the availability of gathered data. Consequently, several comparative studies and assessments of several approaches have been accompanied (Lu et al., 2005; Chen et al., 2005; Notaro, 2008).

In order to determine the optimum way to compute PET for use in rainfall-runoff models, Oudin et al. (2005), Tegos et al. (2015), and Mcmahon et al. (2016) examined and evaluated historical advances of PET algorithms using standard meteorological data. According to McMahon et al. (2016), the Penman–Monteith method interpretation was highly effective for evaluating PET.

To calculate the ET, an open pan evaporimeter was installed at Dwali station, but we found that this machine gives either inaccurate or insignificant results which cannot be practiced for the future research purpose; it is also proved that an open pan evaporimeter is not sophisticated enough to quantify the ET in the glacier region of the Himalayas. An attempt has been made to use the meteorological and modeled data to estimate ET and to compute ET and/or PET in the Pindar valley. To quantify the ETo from meteorological parameters, we first set all data; the brief details of the meteorological parameters are presented in Table 14.1.

14.2 STUDY AREA

This study deals with the analysis and quantification of the ETo from different methods and algorithms from two adjacent glacier valleys of the Central Himalayas, namely, Pindar and Kafni. The Pindari and Kafni glaciers lie in the Pindar valley and drain into the Alaknanda River (upper Ganga basin) in the Bagehswar district of Uttarakhand state (Chauhan et al., 2022) (Figure 14.1a). The total length of Pindari Glacier is about ~5.9 km, and it covers an area of 9.6 km^2. However, the length and area of Kafni glacier are 4.21 km and 3.3 km^2, respectively. The valley lies between latitudes 30°12′15″ and 30° 19′ 10″N and longitudes 79°59′00″ and 80°01′55″E. The area of the basin of Pindari Glacier is 111 km^2 and Kafni Glacier sub-basin was delineated ~62 km^2, and whole area combining Pindari and Kafni Glacier basins up to Dwali occupies about ~173 km^2, extended over an altitude range of 2570–6183 m (asl). Moreover, the Pindari glacier is one of the most retreating glaciers in the central Himalayan region. The Pindari and Kafni are the south-facing glaciers. The lithology of the glaciated valley is dominated by Higher Himalayan crystalline rocks belonging to the Pindari formation of the Vaikrita Group (Valdiya, 1999). The terrain has preserved diverse glacio-geomorphological features, both erosional and depositional. These include the U-shaped glacial trough of trunk and tributary glaciers, recessional morainic ridges, hanging valleys, debris fans, kame terraces, and earth pyramids (Bali et al., 2013). An area is indurated, rounded linear ridges, dominated by crystalline lithoclats (gniesses, schists, and calc-silicates rocks). The preservation is poor which we attribute to the postdepositional reworking caused due to the enhanced denudational rates triggered by higher monsoonal precipitation and active tectonic processes (Bali et al., 2011).

TABLE 14.1 Year-Wise (2017–2019) Details of the Meteorological Parameters Used to Quantify the ET in the Pindar River Basin

Period		Temperature (°C)		Relative Humidity (%)		Wind Speed ($m\ s^{-1}$)	SunShine (h)	Net Radiation ($MJ\ m^{-2}$)	Rainfall (mm)	Reference Evapotranspiration (mm)
		T_{max}	T_{min}	RH_{max}	RH_{min}	U_2	n	R_n	P	ET_o
Year	Month	(°C)	(°C)	(%)	(%)	($m\ s^{-1}$)	(h)	($MJ\ m^{-2}$)	(mm)	(mm)
2017	Jan	8.0	3.0	58.2	32.3	0.8	6.5	12.5	44.2	1.1
	Feb	8.2	3.2	62.2	39.2	0.8	7.0	15.6	45.6	1.5
	Mar	11.9	6.1	72.2	43.0	0.8	8.5	19.9	84.5	2.2
	Apr	12.8	8.5	78.2	41.2	0.9	9.5	23.6	135.3	3.1
	May	16.8	7.5	82.2	36.2	0.9	8.0	21.8	109.1	3.3
	Jun	19.0	12.5	86.2	68.2	0.6	7.5	22.1	207.2	3.3
	Jul	19.9	15.2	91.2	73.2	0.5	6.0	18.9	600.0	3.1
	Aug	17.4	14.5	88.2	75.2	0.5	7.0	19.6	613.8	3.1
	Sep	17.5	13.0	89.2	73.2	0.6	6.0	16.5	306.4	2.6
	Oct	12.0	6.0	75.2	52.3	0.8	7.0	15.3	2.4	2.1
	Nov	9.2	3.7	70.2	40.3	0.8	7.0	12.9	27.4	1.3
	Dec	7.2	2.0	68.2	38.2	0.6	6.0	10.8	21.6	1.1
2018	Jan	8.3	2.8	60.0	35.6	0.8	6.5	12.5	30.4	1.3
	Feb	8.8	4.0	68.6	41.9	0.8	8.0	15.8	69.0	1.8
	Mar	11.9	6.5	70.2	42.9	0.7	8.5	19.9	104.4	2.6
	Apr	13.6	9.0	77.1	39.1	1.0	9.0	22.2	160.0	3.3
	May	17.7	7.8	79.1	38.0	0.8	8.0	21.8	103.4	3.6

TABLE 14.1 *(Continued)*

Period		Temperature (°C)		Relative Humidity (%)		Wind Speed (m s^{-1})	SunShine (h)	Net Radiation (MJ m^{-2})	Rainfall (mm)	Reference Evapotranspiration (mm)
		T_{max}	T_{min}	RH_{max}	RH_{min}	U_2	n	R_n	P	ET_o
	Jun	18.6	13.7	92.1	70.8	0.5	7.5	22.1	349.6	3.6
	Jul	19.6	16.1	85.7	74.6	0.4	7.0	20.4	1024.6	3.7
	Aug	17.7	15.8	89.9	85.7	0.4	7.0	19.6	720.0	3.5
	Sep	18.0	13.5	88.6	67.0	0.7	6.0	16.5	429.2	2.8
	Oct	12.7	6.3	73.5	41.2	0.8	8.0	16.5	53.8	2.4
	Nov	8.8	3.5	79.0	45.1	0.8	7.0	12.9	59.4	1.5
	Dec	7.9	1.4	72.5	41.6	0.5	6.0	10.8	41.6	1.1
2019	Jan	6.8	1.8	57.2	32.3	0.8	6.5	12.5	62.4	1.3
	Feb	7.1	2.3	63.2	39.2	0.8	7.5	15.8	12.2	1.7
	Mar	10.8	5.6	71.2	43.0	0.8	8.5	19.9	21.2	2.5
	Apr	11.9	7.8	76.2	39.2	1.0	9.3	22.2	49.8	3.3
	May	15.2	7.5	82.0	37.1	0.8	8.0	21.8	57.2	3.5
	Jun	17.5	11.2	85.2	65.2	0.6	7.5	22.1	228.8	3.5
	Jul	18.2	14.5	92.2	68.2	0.4	6.5	20.4	537.3	3.4
	Aug	19.2	15.2	89.2	71.2	0.4	7.0	19.6	700.0	3.4
	Sep	16.9	14.2	90.2	69.2	0.6	6.0	16.5	259.6	2.9
	Oct	13.5	6.2	78.2	50.2	0.8	7.5	16.5	18.0	2.4
	Nov	8.2	3.0	65.2	39.2	0.8	7.0	12.9	41.0	1.6
	Dec	6.2	2.0	63.2	37.2	0.6	6.0	10.8	21.0	1.1

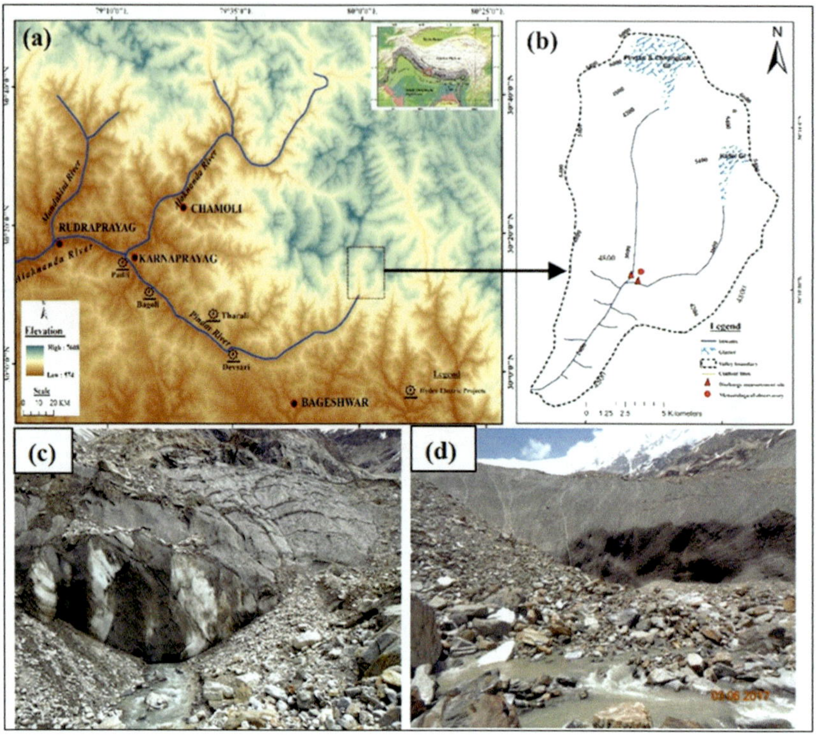

FIGURE 14.1 Study area map (a) shows the location of the study area in central Himalaya, (b) discharge sites and streams from the two adjacent glacier valleys, (c) snout of the Pindari glacier, and (d) snout point of the Kafni glacier in ablation season.

14.3 DATA AND METHODOLOGY

To quantify the ETo, continuous (2017–2019) daily *in situ* meteorological data was collected. The estimated PET for each year was derived from the initial Penman–Monteith equation, as well as the equations of aerodynamic and canopy resistance, which are derived in Chapter 2 in the FAO Penman–Monteith Equation (Allen et al., 1998), as given below in Equation 14.1:

$$ET_o = \frac{0.408\Delta(R_n - G) + \gamma \dfrac{900}{T+273} u_2(e_s - e_a)}{\Delta + \gamma(1 + 0.34u_2)} \tag{14.1}$$

where ET_0 refers to reference evapotranspiration (mm day^{-1}), R_n refers net radiation at the crop surface (MJ m^{-2} day^{-1}), G refers soil heat flux density

(MJ m^{-2} day^{-1}), T refers air temperature at 2 m height (°C), u_2 refers wind speed at 2-m height (m s^{-1}), e_s refers saturation vapor pressure (kPa), e_a refers actual vapor pressure (kPa), $e_s - e_a$ refers saturation vapor pressure deficit (kPa), Δ refers slope vapor pressure curve (kPa $°C^{-1}$), and γ refers psychrometric constant (kPa $°C^{-1}$).

14.4 RESULTS AND DISCUSSION

The climate of the central Himalayan region is considerably influenced by the southwest monsoon between May and October. The data was collected from the rainfall (2017–2019) at the meteorological observatory at Dwali station in the Pindari Kafni glacier valleys. It shows that out of all rainfall events, maximum rainfall occurs with an amount of 788.8 mm (55.15%) in the month of July and a minimum rainfall was computed 101.6 mm (3.80%) in the month of October. An average rainfall of 413.1 mm is quantified during the studied period (May–October) (Table 14.1).

Three-year (2017–2019) total averaged rainfall exhibits 2106.58 ± 119.21 (±SE (standard error)) mm significant variations from year to year in the studied period. Total rainfall received in 2017–2019 was 1838.90, 2680, and 1800.85 mm, respectively (Chauhan et al., 2022). Maximum rainfall was received in year 2018, as years 2017 and 2019 received almost the same amount of rainfall in the study area. The mean monthly average rainfall for the study period (2017–2019) was estimated to be 351.10 mm. The average daily maximum and minimum temperatures in the studied period (2017–2019) were computed 17.31°C and 11.17°C, with an average mean temperature of 14.71 ±1.05°C (±SE) (Table 14.1). The distribution of temperature over the studied period (2017–2019) shows that the mean monthly average temperatures from May to October were 11.90, 15.30, 17.31, 16.92, 15.64, and 11.17°C, respectively. The daily values of relative humidity ranged between 59.59 and 87.78%. The mean monthly relative humidity for May–October (2017–2019) was estimated to be 73.14 ±5.03% (±SE) and details are illustrated in Table 14.1. The ablation period year wise (2017–2019) total interannual precipitation trend can be seen in Figure 14.2. The detail temperature and rainfall correlation trends are presented with discharge and sediment in the study.

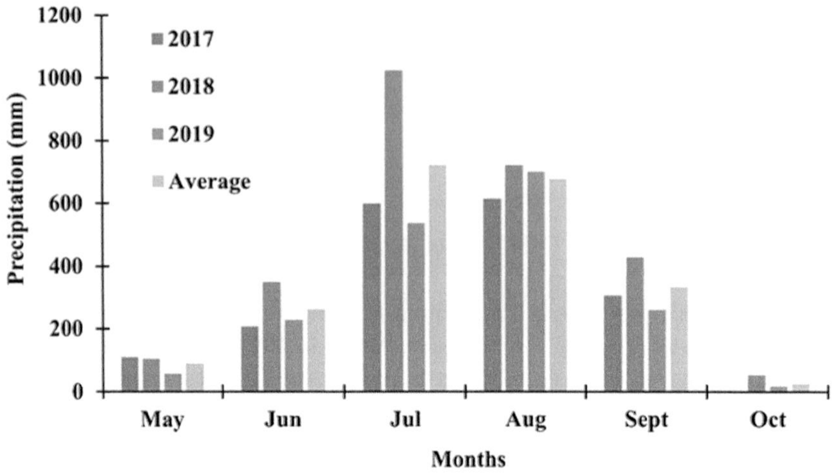

FIGURE 14.2 Monthly precipitation in mm per day in the study area.

Three-year ETo based on ET was quantified for the years (2017–2019) based on *in situ* hydro-meteorological observation data. The annual ETo observed in the Pindari valley was 2.49 mm day^{-1} (SD±0.39). Furthermore, the results were presented seasons wise (Figure 14.1 and Table 14.1). Results show that maximum ETo was computed (3.41 mm day^{-1} ±0.07) for the summer season and minimum ETo was observed (1.35 ±0.30 mm day^{-1}) in the winter season. Spring and autumn season ETo was observed (3.04 ±0.55 mm day^{-1}) and (2.17 ± 0.65 mm day^{-1}), respectively. The details of the meteorological parameters to derive the ETo for the study area are presented in Table 14.1. In the study, ETo was computed from two standard methods: (1) FAO method and (2) Cropwat 8.0, and both method results were compared and validated from modeled Global Land Cover Facility Modis data (Figure 14.3). Results show variability (Figure 14.4) from different methods and observation, which resulted that the patterns of the ETo from both applied methods and modeled Modis data are the same in months in three years (2017–2019) average.

In the study area, the average three-year (2017–2019) seasonal ETo was also quantified from the interannual in the Pindari glacier river. It was observed that maximum ETo to be computed (34%) in the summer season and minimum (14%) the winter season. Rest of the other seasons, i.e., Spring and Autumn ETo quantified (30%) and (22%), respectively, in the valley. Season-wise details of the ETo are presented in Figure 14.5.

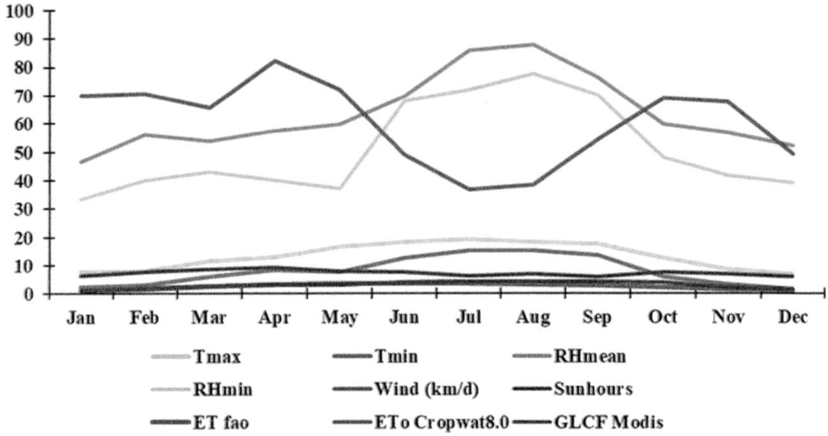

FIGURE 14.3 Three-year average meteorological parameters quantified for ETo and two standard methods and modeled Modis data are presented monthly.

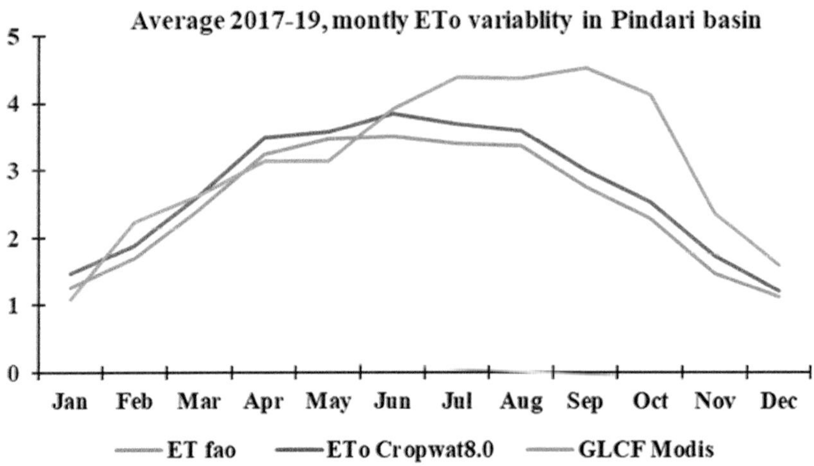

FIGURE 14.4 Pattern and behaviors of the ETo in the Pindar valley from different methods.

14.5 CONCLUSION

In conclusion, observations to quantify the ETo are in nutshell:

- ETo to be calculated from meteorological data, using Penman–Monteith equation.

- ETo was measured from Cropwat and FAO and observed the same pattern
- Average study period and monthly ETo were observed in the months of June and July
- Observed ET was validated with MODIS reanalysis data and achieved the value $R^2 = 0.70$ and 0.56 FAO and Cropwat, respectively, which shows a good correlation.
- Season-wise maximum ETo is calculated to be in summer and minimum in winter.
- Humidity and rainfall are the key factors to loss water in hydrological cycle analysis.

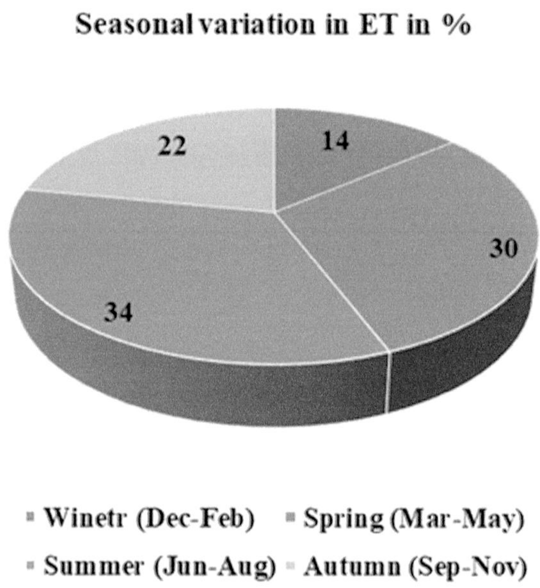

FIGURE 14.5 Three-years (2017–2019) seasonal ETo variations in percentage for the Winter, Summer, Spring, and Autumn months.

DATA AND MATERIAL AVAILABILITY

Data needed to evaluate the conclusions are published and presented in the chapter and the Supplementary Material. Any additional data/code related to this chapter may be requested from the authors.

ACKNOWLEDGMENTS

The first author thankfully acknowledges Science and Engineering Research Board/Department of Science and Technology for the financial support for this research project (File No. EEQ/2016/000292). The authors are thankful to the Director, Wadia Institute of Himalayan Geology, for all logistical support.

COMPETING INTERESTS

All authors declare no competing interests.

KEYWORDS

- **Evapotranspiration**
- **Pindari valley**
- **meteorological parameters**
- **Kafni glacier**
- **Penman–Monteith equation**

REFERENCES

Allen, R. G., Pereira, L. S., Raes, D., and Smith, M. (1998). Chapter 2 FAO Penman-Monteith equation. Crop evapotranspiration—Guidelines for computing crop water requirements. FAO Irrigation and Drainage Paper, 56.

Allen, R. G., Masahiro, T., and Ricardo, T. (2007). Satellite-based energy balance for mapping evapotranspiration with internalized calibration (METRIC)—Model. *Journal of Irrigation and Drainage Engineering 133*, 380–394.

Bali, R., Agarwal, K. K., Nawaz, A. S., and Srivastava, P. (2011) Is the recessional pattern of Himalayan glaciers suggestive of anthropogenically induced global warming? *Arabian Journal of Geosciences 4*, 1087–1093.

Bali, R., Nawaz, A. S., Agarwal, K. K., Rastogi, S. K., Krishna, K., and Srivastava, P. (2013) Chronology of late quaternary glaciation in the Pindari valley, Alaknanda basin, central Himalaya (India). *Journal of Asian Earth Sciences 66*, 224–233.

Chauhan, P., Akıner, M. E., Sain, K., and Kumar, A. (2022). Forecasting of suspended sediment concentration in the Pindari-Kafni glacier valley in Central Himalayan region considering the impact of precipitation: Using soft computing approach. *Arabian Journal of Geosciences 15*(8), 1–18.

Chen, D., Gao, G., Xu, C., Guo, J., and Ren, G. (2005). Comparison of the Thornthwaite method and pan data with the standard Penman–Monteith estimates of reference evapotranspiration in China. *Climate Research 28*, 123–132.

Dinpashoh, Y., Jhajharia, D., Fakheri-Fard, A., Singh, V. P., and Kahya, E. (2011). Trends in reference crop evapotranspiration over Iran. *Journal of Hydrology 399*, 422–433.

Khanmohammadi, N., Rezaie, H., Montaseri, M., and Behmanesh, J. (2018). The application of multiple linear regression method in reference evapotranspiration trend calculation. *Stochastic Environmental Research and Risk Assessment 32*, 661–673.

Kousari, M. R., and Ahani, H. (2012). An investigation on reference crop evapotranspiration trend from 1975 to 2005 in Iran. *International Journal of Climatology*, *32*, 2387–2402.

Liu, B., Li, Y., Chen, J., and Chen, X. (2016). Long-term change in precipitation structure over the karst area of Southwest China. *International Journal of Climatology 36*, 2417–2434.

Lu, J., Sun, G., Mcnulty, S. G., and Amatya, D. M. (2005). A comparison of six potential evapotranspiration methods for regional use in the Southeastern United States. *Journal of the American Water Resources Association 41*, 621–633.

McMahon, T. A., Finlayson, B. L., and Peel, M. C. (2016). Historical developments of models for estimating evaporation using standard meteorological data. *Wiley Interdisciplinary Reviews: Water 3*, 788–818.

Notaro, M. (2008). Statistical identification of global hot spots in soil moisture feedbacks among IPCC AR4 models. *Journal of Geophysical Research: Atmospheres 113*(D9).

Oudin, L., Hervieu, F., Michel, C., Perrin, C., Andréassian, V., Anctil, F., and Loumagne, C. (2005). Which potential evapotranspiration input for a lumped rainfall–runoff model? Part 2—Towards a simple and efficient potential evapotranspiration model for rainfall–runoff modelling. *Journal of Hydrology 303*, 290–306.

Papaioannou, G., Kitsara, G., and Athanasatos, S. (2011). Impact of global dimming and brightening on reference evapotranspiration in Greece. *Journal of Geophysical Research 116*, D09107, DOI:10.1029/2010JD015525.

Senay, G. B., Budde, M., Verdin, J. P., and Melesse, A. M. (2007). A coupled remote sensing and simplified surface energy balance approach to estimate actual evapotranspiration from irrigated fields. *Sensors 7*, 979–1000.

Tegos, A., Malamos, N., and Koutsoyiannis, D. (2015). A parsimonious regional parametric evapotranspiration model based on a simplification of the Penman–Monteith formula. *Journal of Hydrology 524*, 708–717.

Valdiya, K. S. (1999). Lithological sub-divisions and tectonics of the Central Crystalline Zone of Kumaon Himalaya. In: *Proceedings of the seminar on geodynamics of Himalayan Region,* Abstr., NGRI, Hyderabad, pp. 204–205

Yao, N., Li, Y., and Sun, C. (2018). Effects of changing climate on reference crop evapotranspiration over 1961–2013 in Xinjiang, China. *Theoretical and Applied Climatology 131*, 349–362.

Yin, Y., Wu, S., Chen, G., and Dai, E. (2010a). Attribution analyses of potential evapotranspiration changes in China since the 1960s. *Theoretical and Applied Climatology 101*(1–2), 19–28.

Yin, Y., Wu, S., and Dai, E. (2010b). Determining factors in potential evapotranspiration changes over China in the period 1971–2008. *Chinese Science Bulletin 55*(29), 3329–3337.

Zhao, J., Xu, Z., Zuo, D., and Wang, X. (2015). Temporal variations of reference evapotranspiration and its sensitivity to meteorological factors in Heihe River Basin, China. *Water Science and Engineering 8*, 1–8.

INDEX

K

L

M

N

O

P

Q

R

S

For Product Safety Concerns and Information please contact our
EU representative GPSR@taylorandfrancis.com Taylor & Francis
Verlag GmbH, Kaufingerstraße 24, 80331 München, Germany